10th International Conference on Computer Simulation in Physics and Beyond (ICCEE 2017)

Journal of Physics: Conference Series Volume 933

Edmonton, Canada
11 – 13 October 2017

ISBN: 978-1-5108-5515-1
ISSN: 1742-6588

Printed from e-media with permission by:

Curran Associates, Inc.
57 Morehouse Lane
Red Hook, NY 12571

Some format issues inherent in the e-media version may also appear in this print version.

Printed by Curran Associates, Inc. (2018)

For permission requests, please contact the Institute of Physics
at the address below.

Institute of Physics
Dirac House, Temple Back
Bristol BS1 6BE UK

Phone: 44 1 17 929 7481
Fax: 44 1 17 920 0979

techtracking@iop.org

Additional copies of this publication are available from:

Curran Associates, Inc.
57 Morehouse Lane
Red Hook, NY 12571 USA
Phone: 845-758-0400
Fax: 845-758-2633
Email: curran@proceedings.com
Web: www.proceedings.com

Table of contents

Volume 933

10th International Conference on Computer and Electrical Engineering

11–13 October 2017, Edmonton, Canada

Accepted papers received: 23 November 2017
Published online: 3 January 2018

Preface

Papers

Image processing and Application

Network security

Data analysis and information network

IOP Conf. Series: Journal of Physics: Conf. Series **933** (2017) 011001 doi:10.1088/1742-6596/933/1/011001

PREFACE

It is our great pleasure to welcome you to 2017 10th International Conference on Computer and Electrical Engineering which will take place in University of Alberta, Edmonton, Canada during October 11-13, 2017. ICCEE2017 is dedicated to issues related to Computer and Electrical Engineering.

The major goal and feature of the conference is to bring academic scientists, engineers, industry researchers together to exchange and share their experiences and research results, and discuss the practical challenges encountered and the solutions adopted. Professors from Canada and Luxembourg are invited to deliver plenary speeches regarding latest information in their respective expertise areas. It will be a golden opportunity for the students, researchers and engineers to interact with the experts and specialists to get their advice or consultation on technical matters, sales and marketing strategies.

These proceedings present a selection from papers submitted to the conference from universities, research institutes and industries. All of the papers were subjected to peer-review by conference committee members and international reviewers. The papers selected depended on their quality and their relevancy to the conference. The volume tends to present to the readers the recent advances in the field of Computer and Electrical Engineering and various related areas, such as Computer Vision, Graphics and Intelligence, Computational and Artificial Intelligence, Advanced Power Semiconductors, Analogue and Digital Signal Processing, Algorithm, etc.

We would like to thank all the authors who have contributed to this volume and also to the reviewers, speakers, chair persons and all the conference participants for their support to ICCEE 2017.

Best Regards!

Yours sincerely,

Conference Organizing Committee
October 20, 2017

IOP Conf. Series: Journal of Physics: Conf. Series **933** (2017) 011002 doi:10.1088/1742-6596/933/1/011002

Conference Committee

Conference Chairs

Prof. Elmar Bollin, Offenburg University of Applied Sciences, Germany
Prof. Peter Plapper, University of Luxembourg, Luxembourg
Prof. Mohamed Al-Hussein, University of Alberta, Canada

Local Chair

Assis. Prof. Rafiq Ahmad, University of Alberta, Canada

Program Chairs

Prof. Mahdi Tavakoli, University of Alberta, Canada
Prof. Yongsheng Ma, University of Alberta, Canada
Prof.Viacheslav Pshikhopov, Southern Federal University, Russian Federation
Prof. Rolf Roskam, Ostfalia University of Applied Sciences, Germany

Publicity Chairs

Prof. Takashi Harada, Kindai University, Japan
Asst. Prof. Hassan HosseinNia, TU Delft/3mE/Precision and Microsystems Engineering (PME), Netherlands
Prof. Sathans Suhag, National Institute of Technology Kurukshetra, India

Technical Committee

Prof. Basel Alsayyed Ahmad, University of Alberta, Canada
Asst. Prof. Dr. Hossein Rouhani, University of Alberta, Canada
Asst. Prof. Dr. Ahmed Qureshi, University of Alberta, Canada
Dr. Sattar J Aboud, Science and Technology of the Iraqi Council of Representatives, Iraq
Dr. Ugur Bilge, Akdeniz Üniversity Faculty of Medicine, Turkey
Prof. Said Yahmedi, Laboratory of Mathematical Modeling and Numerical Simulation Annaba, Algeria
Prof. Kamel Eddine HEMSAS, University Ferhat Abbas Setif -1-, Algeria
Prof. Dr.Ing. Rainer Müller, Zentrum für Mechatronik und Automatisierungstechnik, Germany
Dr.Earle Jennings, QSigma, Inc., US
Asst. Prof. Dr. Waheeb Abu-Dawwas, Applied Science Private University, Jordan
Asst. Prof. Dr. Levent Seyfi, Selcuk University , Turkey
Prof. T. Cetin Akinci, Istanbul Technical University, Turkey
Asst. Prof. Dr. Akif Durdu, Selcuk University, Turkey
Prof. Parvin Lakbala, Hormozgan University of Medical Sciences, Iran

IOP Conf. Series: Journal of Physics: Conf. Series **933** (2017) 011003 doi:10.1088/1742-6596/933/1/011003

Peer review statement

All papers published in this volume of *Journal of Physics: Conference Series* have been peer reviewed through processes administered by the proceedings Editors. Reviews were conducted by expert referees to the professional and scientific standards expected of a proceedings journal published by IOP Publishing.

IOP Conf. Series: Journal of Physics: Conf. Series **933** (2017) 012001 doi:10.1088/1742-6596/933/1/012001

Mobile Visual Search Based on Histogram Matching and Zone Weight Learning

Chuang Zhu, Li Tao, Fan Yang, TaoLu, HuizhuJia and Xiaodong Xie

Peking University, No. 5 Yiheyuan Road, Beijing, China

{czhu, hzjia}@pku.edu.cn

Abstract. In this paper, we propose a novel image retrieval algorithm for mobile visual search. At first, a short visual codebook is generated based on the descriptor database to represent the statistical information of the dataset. Then, an accurate local descriptor similarity score is computed by merging the *tf-idf* weighted histogram matching and the weighting strategy in compact descriptors for visual search (CDVS). At last, both the global descriptor matching score and the local descriptor similarity score are summed up to rerank the retrieval results according to the learned zone weights. The results show that the proposed approach outperforms the state-of-the-art image retrieval method in CDVS.

1. Introduction

Content-based image retrieval (CBIR) is always a hot topic to study these years [1]. With the steadily growing amounts of mobile devices, a new type of CBIR technique, mobile visual search [2], is attracting keen attention of researchers. In mobile visual search, there are several significant challenges: limited wireless network bandwidth, small mobile battery capacity, little memory space to store features and the requirement of feature interoperability [3]. To address these challenges, the ISO/IEC moving pictures experts group (MPEG) drafts the compact descriptors for visual search (CDVS) [4].

A typical CDVS application framework is shown in Figure 1. The local features are extracted and compressed to produce compact visual descriptors on the mobile devices and the retrieval is performed on the remote server using the received descriptors. In this paper, we focus on the image retrieval algorithm design.

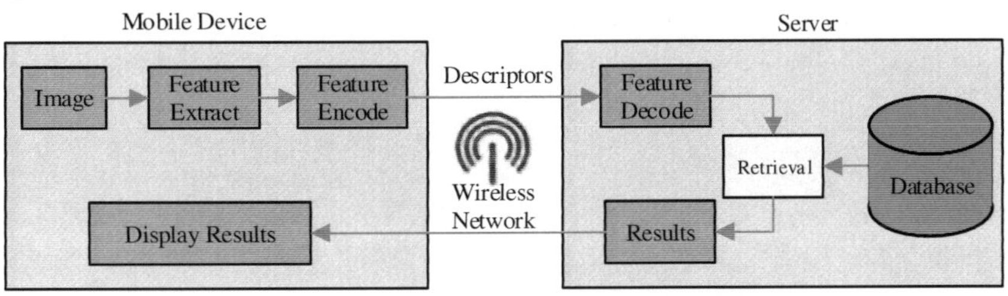

Figure 1. Mobile visual search architecture.

Generally, for an image retrieval system, the user supplies a query image or a query object by selecting a region of the query image, and then the feature detection module extracts local features such as SIFT [5] and SURF [6]; the retrieval system returns a ranked list of images that contain the same object based on the Bag-of-Words (BoW) [7] signature or global descriptors such as fisher vectors (FV) [8] and vector of locally aggregated descriptors (VLAD) [9]. In this paper, the above word "feature" refers to the original uncompressed key point, and the word "descriptor" means the encoded local feature or the aggregated global feature. However, the BoW signature and global descriptors are generated based on the orderless local features, which leads to the disregarding of information about the spatial layout of the features. To further improve image retrieval performance, Philbin et al. [10] propose to add an efficient spatial verification to rerank the results returned from the BoW model. Similarly, in work [11] the authors first retrieve videos using the weighted frequency vector and then rerank them based on the spatial consistency measure. Although the above BoW based approaches can yield decent performance, the fact that mobile visual search generally requires low memory makes these approaches unsuitable because of theirvery large visual word tables [3]. Thus, the CDVS standard adopts the Scalable Compressed Fisher Vector (SCFV) [12] to conduct image retrieval, and then reranks the returned results by using geometric consistency check (GCC) which includes a ratio test and a fast geometric model estimation [3]. However, the statistical information of image database, such as "inverse document frequency" (*idf*), is ignored and the global similarity score is discarded in reranking stage.

In this paper, we propose a novel visual search algorithm. First, a short visual codebook is generated based on the database descriptors to represent the statistical information of the dataset. Then, an accurate local descriptor similarity score (LDSS) is computed by merging the "term frequency-inverse document frequency" (*tf-idf*) weighted histogram matching and the ratio based weighting strategy in CDVS. At last, both the global descriptor matching score (GMS) and the LDSS are summed up to rerank the retrieval results according to the learned zone weights.

The remainder of the paper is structured as follows. In Section 2, we overview the mobile visual search techniques. In Section 3, we present details of our proposed image retrieval algorithm. In Section 4, we discuss the performance comparisons and then in Section 5 we conclude our paper.

2. Mobile visual search overview

The CDVS standard defines the feature extraction process and 6 different query descriptor lengths (0.5KB, 1KB, 2KB, 4KB, 8KB and 16KB) to support different scenarios [3]. The standardized CDVS bitstream makes the interoperability of the descriptors from different devices possible, and CDVS standard provides a matching mechanism for descriptors with different coding rates. For a typical mobile visual search application, there are two stages to perform: CDVS bitstream extraction and image retrieval using CDVS bitstream.

2.1. CDVS Bitstream Extraction

Figure 2 outlines the work flow of the CDVS bitstream extraction, which includes seven building blocks: interest point detection, local feature selection, local feature description, local feature compression, local feature aggregation, local feature location compression and CDVS encoding.

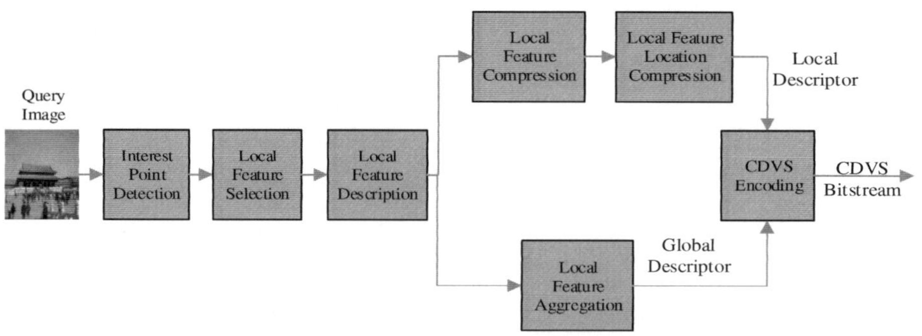

Figure 2. CDVS Bitstream extraction process.

For a query image, the CDVS standard first adopts a low-degree polynomial (ALP) detector [3] to find the interest points by approximating the result of the Laplacian of Gaussian (LoG) filter. Secondly, a subset of local features is selected to meet the bandwidth limitation according to a statistically learned relevance measure, which indicates the priori probability of a feature from query image matching a feature of database image correctly. After local feature selection, each picked local feature is described as original 128-dimensional (1024 bits) SIFT vector [5]. Then, on one hand, CDVS standard adopts a SCFV model to aggregate the local features to build a global descriptor for the query image; on the other hand, CDVS adopts a transform coding scheme followed by a scalar quantization and entropy coding to compress the selected local SIFT features. Besides, the local feature location compression is performed to record the x and y location information, which will be used in the GCC step of image retrieval. At last, in CDVS encoding module, the global descriptor, the compressed local features and the coded locations are merged to produce the CDVS bitstream.

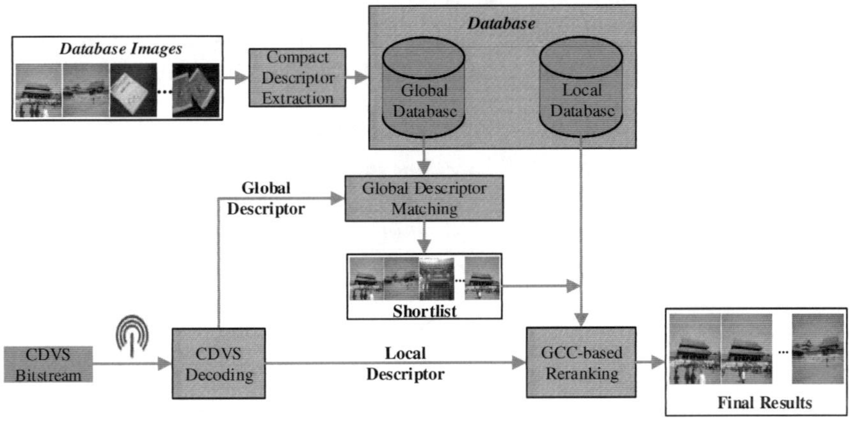

Figure 3. Image retrieval based on CDVS bitstream.

2.2. Image Retrieval Using CDVS Bitstream

The CDVS also specifies the image retrieval framework. With the same extraction flow above, a set of global and local descriptors of the database images are extracted, and then they are indexed to build a global database and a local database, respectively. With the received query image CDVS bitstream, image retrieval will be performed according to the framework in Figure 3.

First, the global and local descriptors are separated out from the query bitstream by the CDVS decoding module. Second, the query global descriptor is compared with each global descriptor in the global database, and based on the GMS a shortlist with the top ranked N images, such as 500, is returned. In the GMS comparison, the Hamming distance-based similarity score S_{GMS} is computed according to

$$S_{GMS}(X,Y)=\frac{\sum_{i=0}^{511} b_i^X b_i^Y w_{Ha(u_i^X,u_i^Y)}(32-2Ha(u_i^X,u_i^Y))}{32\sqrt{\sum_{i=0}^{511} b_i^X}\sqrt{\sum_{i=0}^{511} b_i^Y}} \tag{1}$$

where X and Y are the SCFVs of two images. In CDVS, SCFV is built based on a selected subset of Gaussian components (512 in total) from the Gaussian Mixture Model (GMM). In Equation (1), $b_i^X = 1$ if i_{th} Gaussian is selected, if not then $b_i^X = 0$. $Ha(u_i^X,u_i^Y)$ and $w_{Ha(u_i^X,u_i^Y)}$ are the Hamming distance and the correlation weight of i_{th} Gaussian between X and Y, respectively. Third, an exhaustive pairwise comparison is performed to find all matched local descriptor pairs between the query image and each candidate image in the shortlist. Then, the ratio test and GCC are conducted to remove the wrong matched pairs, which are also called outliers. The left P inlier good match pairs are used to generate the matching score L between two images. At last, the reranked final image retrieval results are generated according to L.

$$L = \sum_{i=1}^{P} \cos(\frac{\pi}{2} \square \frac{min_i}{smin_i}) \quad (2)$$

In Equation (2), *cos* is the cosine transform function, and $min_i/smin_i$ is the ratio between the distance of the closest neighbor and that of the second-closest neighbor in computing the i_{th} inlier matching pair [5].

3. Proposed image retrieval

In this section, we first present our proposed image retrieval architecture, and then show our designed LDSS computing scheme. Finally, we illustrate our zone weight learning (ZWL) based image reranking algorithm, which takes both GMS and LDSS into consideration.

3.1. Image Retrieval Architecture

The main image retrieval flow is the same with the typical CDVS application illustrated in Figure 3: first generate shortlist and then perform image reranking. As shown in Figure 4, the main differences between our proposed image retrieval architecture and the typical CDVS image retrieval system come from the following three aspects.

First: Local database statistical information is integrated into the image retrieval system. The local database is a collection of local descriptors, and we cluster them to create a l-size, such as 300, codebook $C=(c_1,c_2,...,c_l)$ by using k-means algorithm. Based on the codebook, we can quantize and represent the local descriptors of database images as visual words and then compute the *idf* for each visual word. The *idf* is calculated according to Equation (3).

$$idf_{ci} = \log \frac{N}{N_i} \quad (3)$$

where c_i (i from 1 to l) is the i_{th} visual word of the codebook, N is the number of database images and N_i is the number of images containing visual word c_i. Then the *idf* weighting table is

$$W_{idf} = (idf_{c1}, idf_{c2}, ..., idf_{cl}) \quad (4)$$

Both the codebook C and the generated weighting table W_{idf} will be used in the image reranking stage.

Second: More information ispassed to the image reranking module, including the GMS, the pre-calculated *idf* weighting table and the local descriptor codebook. It is noteworthy that the GMS corresponding to each image in the shortlist is generated in the global matching stage.

Third: The adopted image reranking algorithm is different. Besides GCC, we add*tf-idf* weighted histogram matching to compute the LDSS. Moreover, we utilize both the GMS and the LDSS to rerank the shortlist based on the ZWL.

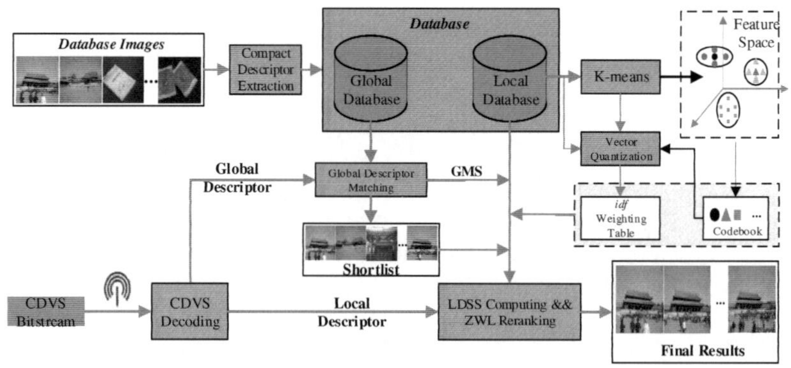

Figure 4. Proposed image retrieval architecture.

3.2. LDSS Computing Based on tf-idf Weighted Histogram Matching

LDSS represents the similarity between the query image local descriptor (QLD) and the reference image local descriptor (RLD) in the database. We use the following Equation (5) to compute LDSS.

$$S_{LDSS} = bK + (1-b)L \tag{5}$$

where K is the *tf-idf* weighted histogram matching score (HMS), L is the reranking criteria used in original CDVS reference model as shown in Equation (2), b is a constant to balance K and L. The key of computing LDSS is to find K.

The LDSS computing flow is shown in Figure 5. We perform local descriptor matching for the QLD and a RLD first, and then conduct GCC to remove the wrongly matched pairs. The left inlier pairs are used to compute K. Finally, we compute LDSS based on K and L.

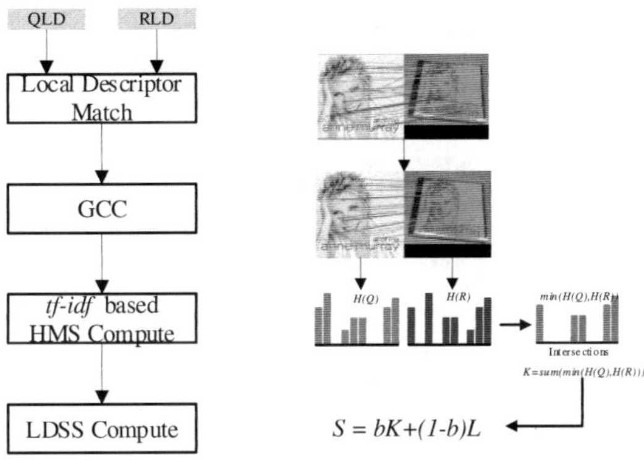

Figure 5. Histogram Matching Based LDSS Computing Flow.

Let QLD and RLD are

$$\begin{cases} Q_{QLD} = (q_1, q_2, ..., q_m) \\ R_{RLD} = (r_1, r_2, ..., r_n) \end{cases} \tag{6}$$

where q_i (i from 1 to m) and r_j (j from 1 to n) are two sets of local descriptors in CDVS standard corresponding to the query image and a database reference image, respectively. After GCC, we get h number of good matching pairs (the inliers) M, as shown in Equation (7).

$$M(Q_{QLD}, R_{RLD}) = \{(q_{i1}, r_{j1}), (q_{i2}, r_{j2}), ..., (q_{ih}, r_{jh})\} \tag{7}$$

where q_{io} and r_{jo} (o from 1 to h) are from Q_{QLD} and R_{RLD}, respectively. Then, q_{io} and r_{jo} are quantized to create two weighting histograms $H(Q)$ and $H(R)$, according to the codebook and *idf* weighting table. $H(Q)$ and $H(R)$ are histograms with l bins, corresponding to the l visual words of the codebook. Based on Equation (4), we have

$$\begin{cases} H(Q) = ((idf_{c1}) \cdot N_{q1}, (idf_{c2}) \cdot N_{q2}, ..., (idf_{cl}) \cdot N_{ql}) \\ H(R) = ((idf_{c1}) \cdot N_{r1}, (idf_{c2}) \cdot N_{r2}, ..., (idf_{cl}) \cdot N_{rl}) \end{cases} \tag{8}$$

where N_{qi} and N_{ri} (i from 1 to l) denote the counts of query and reference inlier points, contained in Equation (7), which fall into the i_{th} bins of $H(Q)$ and $H(R)$, respectively.

We know that histogram intersection effectively counts the number of points in two sets that fall into the same bin, and we use this criteria to evaluate the matching degree of $H(Q)$ and $H(R)$. Then, we get the *tf-idf* weighted histogram matching score K

$$K(H(Q), H(R)) = \sum_{i=1}^{l} \min(H(Q_i), H(R_i)) \tag{9}$$

where *min* is the minimum function.

Combining Equation (2) and (9), we rewrite Equation (5) as

$$S_{LDSS} = bK + (1-b)L$$
$$= b\sum_{i=1}^{l} \min(H(Q_i), H(R_i)) + (1-b)\sum_{i=1}^{P} \cos(\frac{\pi}{2}\Box\frac{min_i}{smin_i}) \quad (10)$$

In this paper, we set b=0.5 throughout experimentsto achieve good balance between histogram matching score *K* and CDVS reranking criteria *L*.

3.3. Image Reranking Based on Zone Weight Learning

Generally, image retrieval first sorts the database images based on BoW [7] or global descriptor [12] and yields a shortlist with top ranked images. Then image reranking algorithm will be followed to refine the results based on just local descriptor matching. In this paper, we propose to perform image reranking by using both LDSS and GMS.

Our motivation is that if two images containing the same targets, they should look similar both from the whole and the part at the same time. In statistical text retrieval system [13], a document can be split into different zones, such as *title*, *abstract* and *body*. In each zone, the document is viewed as a sequence of terms. We usually conduct text retrieval by combining the information of different zones. For example, we often retrieve a document like this: find documents with "computer" in the *title* and "adam" in the *author* list and the phrase "system architecture" in the *body*. The matching scores of different zones are weighted and added together to sort the documents. Similarly, in this paper we treat the global information of an image as *global* zone (similar to the *title* or *abstract* zones in text retrieval) and local information as *local* zone (similar to the *body* zone in text retrieval). We take both *global* zone and *local* zone matching scores into consideration for image reranking.

We propose to use the following equation in image reranking.

$$S = \lambda S_{GMS} + (1-\lambda) S_{LDSS} \quad (11)$$

where S_{GMS} is the GMS presented in Equation (1) and $S_{LDSS}$is LDSS illustrated in Equation (10). In Equation (11), $\lambda$and (1-λ) are the weights for *global* zone and *local* zone, respectively. In the following, we will describe the proposed zone weight learning method to findλ.

Firstly, the image dataset is divided into a training set and a testing set, and the training set is used to learn zone weights. Then, for the training set, we match each image with the others to generate a series of image pairs, in which the image pairs containing the same targets are labelled as α and the others arelabelled asβ. At last, we select *P* image pairs with labelα and *N* image pairs with labelβ to trainλas follows. We first normalize S_{GMS} and S_{LDSS} of the pairs to a range of [0, 1] by using min-max normalization, as shown in Equation (12).

$$y = (x - MIN)/(MAX - MIN) \quad (12)$$

where x is the input value and y is the normalized output. *MIN* and *MAX* are the minimum and maximum of the input values.Then, let $Pa_\alpha$and Pa_βas

$$\begin{cases} Pa_\alpha = \{(S'_{GMS\alpha1}, S'_{LDSS\alpha1}),...,(S'_{GMS\alpha i}, S'_{LDSS\alpha i}),(S'_{GMS\alpha P}, S'_{LDSS\alpha P})\} \\ Pa_\beta = \{(S'_{GMS\beta1}, S'_{LDSS\beta1}),...,(S'_{GMS\beta j}, S'_{LDSS\beta j}),(S'_{GMS\beta N}, S'_{LDSS\beta N})\} \end{cases} \quad (13)$$

where S'_{GMS_i} and S'_{LDSS_i} are the i_{th} normalized GMS and LDSS of the image pairs. For explanation convenience, we randomly select 300pairs with labelα and 300pairs withlabelβ, and depict the corresponding reranking scores *S* with differentλvalues, as shown in Figure 6.

We can see that differentλvalues generate different distributions of *S*. Then, the question becomes how to choose a properλwhich generates an*S* distribution that can be easily classified by a fixed threshold value. We propose to use Equation (14) to trainλ.

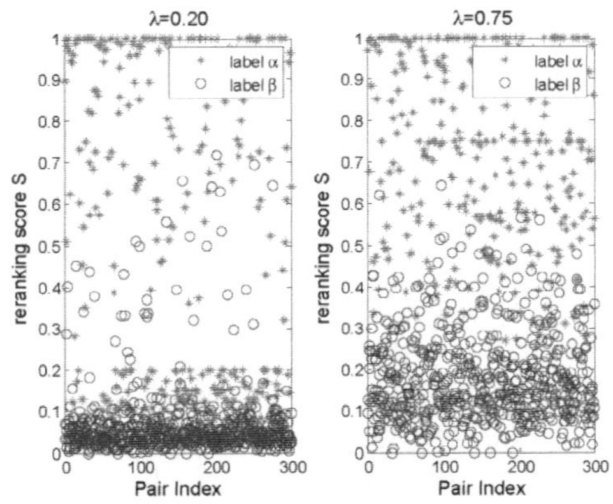

Figure 6. The distribution of reranking score with different λ.

$$\hat{\lambda} = \underset{\lambda}{\operatorname{argmin}} = \frac{C_{inner}}{D_{inter}} \quad (14)$$

where C_{inner} represents the sum of concentration degree in each class, D_{inter} denotes the separation degree between classα and classβ. We will formulate Equation (14) in detail in the following.Corresponding to $Pa_\alpha$and Pa_β, we have

$$\begin{cases} S'_\alpha(\lambda) = \{S'_{\alpha 1}(\lambda), \ldots, S'_{\alpha i}(\lambda), \ldots, S'_{\alpha P}(\lambda)\} \\ S'_\beta(\lambda) = \{S'_{\beta 1}(\lambda), \ldots, S'_{\beta j}(\lambda), \ldots, S'_{\beta N}(\lambda)\} \end{cases} \quad (15)$$

with

$$\begin{cases} S'_{\alpha i}(\lambda) = \lambda S'_{GMS\alpha i} + (1-\lambda) S'_{LDSS\alpha i} \\ S'_{\beta j}(\lambda) = \lambda S'_{GMS\beta j} + (1-\lambda) S'_{LDSS\beta j} \end{cases} \quad (16)$$

where i from 1 to P and j from 1 to N. Let $m_\alpha(\lambda)$ and $m_\beta(\lambda)$ are the expectations, and $var_\alpha(\lambda)$ and $var_\beta(\lambda)$ are the variances of $S'_\alpha(\lambda)$ and $S'_\beta(\lambda)$, respectively. Then we can get the expectation of the whole, including both $S'_\alpha(\lambda)$ and $S'_\beta(\lambda)$.

$$m'(\lambda) = R_\alpha m_\alpha(\lambda) + R_\beta m_\beta(\lambda) \quad (17)$$

where $R_\alpha = P/(P+N)$ and $R_\beta = N/(P+N)$. Finally, we yield C_{inner} and D_{inter} as

$$\begin{cases} C_{inner} = R_\alpha(var_\alpha(\lambda)) + R_\beta(var_\beta(\lambda)) \\ D_{inter} = R_\alpha(m_\alpha(\lambda) - m'(\lambda))^2 + R_\beta(m_\beta(\lambda) - m'(\lambda))^2 \end{cases} \quad (18)$$

We substitute Equation (18) into Equation (14) producing

$$\hat{\lambda} = \underset{\lambda}{\operatorname{argmin}} \frac{R_\alpha(var_\alpha(\lambda)) + R_\beta(var_\beta(\lambda))}{R_\alpha(m_\alpha(\lambda) - m'(\lambda))^2 + R_\beta(m_\beta(\lambda) - m'(\lambda))^2} \quad (19)$$

Our proposed zone weight learning method will train the bestλby minimizing Equation (19). One simple solution forλis directly checking all the results according to Equation (19) by increasingλwith a small step, such as 0.01, from 0 to 1. Once we obtain the bestλ, we will rerank the shortlist according to

$$S'(\hat{\lambda}) = \hat{\lambda} S'_{GMS} + (1-\hat{\lambda}) S'_{LDSS} \quad (20)$$

whereS'_{GMS} and S'_{LDSS} are the normalized values, and $\hat{\lambda}$ is the best λ.

4. Experiments

Dataset. We evaluate our proposed method on two benchmark datasets: INRIA Holidays [14] and University of Kentucky Benchmark (UKBench) [10]. INRIA Holidays dataset includes 1491 images, and 500 ground truth queries can be used for testing. UKBench dataset contains 2550 tagged ground truth groups, and each group contains 4 pictures of the same object with different views.

Performance evaluation criteria. The mean average precision (mAP) is used to evaluate search performance. The mAP is shown in Equation (21).

$$mAP = \frac{1}{N}\sum_{i=1}^{N}\frac{\sum_{r=1}^{M^i_{relevant}} P(r)}{M^i_{relevant}} \quad (21)$$

where$i$ is the query image index and N is the number of total queries. $M^i_{relevant}$ is the number of relevant images corresponding to the i_{th} query, r is the relevant image index and $P(r)$ is the precision at the cut-off rank of r_{th} relevant image.

Zone weight learning. For each selected dataset, we divide it into two parts. 100 images and 300 images are used in zone weight λ training for INRIA Holidays and UKBench, respectively, and the rest part is used for the visual search performance evaluation. In CDVS, there are six specified bitrate modes (from 0.5KB to 16KB), thus we learn six different weights for them. Figure 7 depicts the learned λ for the two datasets. With the increase of bitrate, more local features are encoded into the CDVS bitstream and thus the reliability of S'_{LDSS} becomes strong, which results in the decrease of λ.

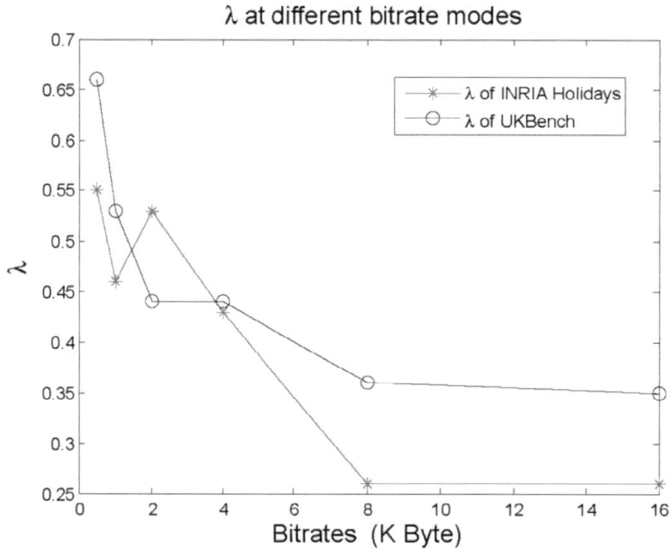

Figure 7. Learned λ in different bitrate modes for INRIA Holidays and UKBench.

Comparisons. For mobile visual search, CDVS achieves the state-of-the-art performance. The image reranking method based on spatial verification in [10] and the image retrieval algorithm in CDVS are tested to make comparisons with our work. All the methods are integrated into CDVS reference software test model framework 11 (TM 11.0) [15]. Test results are tabulated in Table 1. The results show that our proposed method achieves the best searching performance among all 3 methods: it outperforms both the method in work [10] and the state-of-the-art method in CDVS reference model TM 11.0 among all bitrate modes. Our method can achieve maximum up to 3% gain in some specific bitrate modes, such as 1KB mode for UKBench, compared to the other algorithms.

Table 1. Performance comparisons.

Dataset	Bitrate (KB)	mAP		
		work [10]	CDVS	Proposed
INRIA Holidays	0.5	59.39	59.42	**59.77**
	1	59.38	60.93	**61.29**
	2	64.52	66.60	**67.18**
	4	66.18	68.62	**69.03**
	8	69.20	70.43	**70.55**
	16	68.95	70.19	**70.31**
UKBench	0.5	76.13	76.24	**76.74**
	1	76.11	77.45	**79.36**
	2	80.76	82.12	**82.74**
	4	82.47	83.56	**84.73**
	8	84.28	84.50	**84.99**
	16	84.28	84.48	**84.97**

5. Conclusion

In this paper, we propose an image retrieval algorithm for mobile visual search. First, a short visual codebook is generated based on the database descriptors and then an accurate LDSS is computed by merging the *tf-idf* weighted histogram matching and the weighting strategy in CDVS. At last, both the GMS and the LDSS are summed up to rerank the retrieval results according to the learned zone weights. The results show that the proposed approach outperforms the state-of-the-art performance of CDVS. Because most part of our proposed method can be pre-computed offline, the online image retrieval computing complexity is just increased slightly compared with CDVS. In the future, we will focus on low complexity image retrieval algorithm design.

6. References

[1] A referenceThis reference has two entries but the second one is not numbered (it uses the 'Reference (no number)' style. Singhai N, Shandilya S K. 2010. A survey on: content based image retrieval systems. *International Journal of Computer Applications.* 4(2), (2010), 22-26.

[2] Girod B, Chandrasekhar V, Chen D M, et al. 2011. Mobile visual search. *IEEE signal processing magazine.* 28(4), (2011), 61-76.

[3] Duan L Y, Chandrasekhar V, Chen J, et al. 2016. Overview of the MPEG-CDVS Standard. *IEEE Transactions on Image Processing.* 25(1), (2016), 179-194.

[4] S. Paschalakis et al. 2015. Information Technology-Multimedia Content Description Interface-Part 13: Compact Descriptors for Visual Search.(2015), ISO/IEC 15938-13.

[5] Lowe D G. 2004. Distinctive image features from scale-invariant keypoints. *International journal of computer vision.* (2004), 60(2), 91-110.

[6] Bay H, Ess A, Tuytelaars T, et al. 2008. Speeded-up robust features (SURF). *Computer vision and image understanding.* (2008), 110(3), 346-359.

[7] Zhang Y, Jin R, Zhou Z H. 2010. Understanding bag-of-words model: a statistical framework. *International Journal of Machine Learning and Cybernetics*.(2010), 1, 43-52.

[8] F. Perronnin, Y. Liu, J. Sanchez, and H. Poirier. 2010. Large-scale image retrieval with compressed Fisher vectors. In: *Proc. IEEE Conf. Comput. Vis. Pattern Recognit. (CVPR)*, (2010), pp. 3384-3391. San Francisco.

[9] H. Jégou, M. Douze, C. Schmid, and P. Perez. 2010. Aggregating local descriptors into a compact image representation. In: *Proc. IEEE Conf. Comput. Vis. Pattern Recognit. (CVPR)*,(2010), pp. 3304-3311. San Francisco.

[10] Philbin J, Chum O, Isard M, et al. Object retrieval with large vocabularies and fast spatial matching. In: *IEEE Conference on Computer Vision and Pattern Recognition,*

CVPR'07,(2007), pp: 1-8.

[11] Sivic J, Zisserman A. 2009. Efficient visual search of videos cast as text retrieval. *IEEE transactions on pattern analysis and machine intelligence.* (2009), 31(4), 591-606.

[12] J. Lin, L.-Y. Duan, Y. Huang, S. Luo, T. Huang, and W. Gao. 2014. Rate- adaptive compact Fisher codes for mobile visual search. *IEEE Signal Process. Lett.* (2014), 21(2), 195-198.

[13] Manning C D, Raghavan P, Schütze H. 2008. Introduction to information retrieval. Cambridge university press, (2008), Cambridge.

[14] Herve Jegou, Matthijs Douze and Cordelia Schmid. 2008. Hamming Embedding and Weak geometry consistency for large scale image search. In: *proceedings of the 10th European conference on Computer vision.* (2008).

[15] Test Model 11: Evaluation framework for Compact Descriptor for Visual Search, (2014), document ISO/IECJTC1/SC29/WG11/N14680.

Acknowledgments

This work was partially supported by grant from the China Postdoctoral Science Foundation under contract No. 2016M590020, the National Science Foundation of China (61421062 and 61602011), the National High Technology Research and Development Program of China (863 Program) under contract No.2015AA015903， the Major National Scientific Instrument and Equipment Development Project of China under contract No. 2013YQ030967, and the National Key Research and Development Program of China under contract No. 2016YFB0401904 and 2016YFB0402001.

IOP Conf. Series: Journal of Physics: Conf. Series **933** (2017) 012002 doi:10.1088/1742-6596/933/1/012002

Comparative implementation of Handwritten and Machine written Gurmukhi text utilizing appropriate parameters

Jaswinder Kaur[1], Dr. Gagandeep Jagdev[2]

[1]Guru Kashi University,Talwandi Sabo,Punjab, India.
[2]Faculty, Dept. of Comp. Sc., Punjabi University Guru Kashi College, Damdama Sahib,Bathinda, Punjab,India.

Corresponding author. Tel.:+91-9501041900; email: drgagan137@pbi.ac.in

Abstract. Optical character recognition is concerned with the recognition of optically processed characters. The recognition is done offline after the writing or printing has been completed, unlike online recognition where the computer has to recognize the characters instantly as they are drawn. The performance of character recognition depends upon the quality of scanned documents. The preprocessing steps are used for removing low-frequency background noise and normalizing the intensity of individual scanned documents. Several filters are used for reducing certain image details and enabling an easier or faster evaluation. The primary aim of the research work is to recognize handwritten and machine written characters and differentiate them. The language opted for the research work is Punjabi Gurmukhi and tool utilized is Matlab.

1. Introduction

Optical Character Recognition is the procedure via which text within an image is converted into the text itself.It has always been a dream that human functions are replicated by a machine. However, during last four decades, this dream has grown to reality. Optical character recognition is considered among prominent applications of pattern recognition. Several commercial systems based on OCR are available in the market today, but still, human reading capabilities cannot be matched [1].In simple terms, OCR is processing via which input in the form of text characters is provided to a computer through an image. OCR has empowered scanned documents to be much more than image files, converting into searchable documents with text content that is computer recognizable. OCR allows people to convert important documents into electronic documents without any need to retype them manually. The accuracy level obtained is good and information processing in done in minimum time [3], [11]. The OCR algorithm is primarily based on the set of learned characters. The characters in the learned set are compared with the scanned image file. The learned set can be generated easily. The focus of the research work in the paper is on differentiating printed and handwritten Gurmukhi text.Deciding someone's scribbled handwriting is much more difficult as compared to recognizing the characters that constitute a neat laser-printed computer text. This is where human brains beat smart computers hands-down. OCR simplifies the process where computers are supposed to recognize text written in image.[3].

2. Gurmukhi script

Gurmukhi characters come into use for writing the Punjabi language. The second Sikh Guru, Shri Guru Angad Dev Ji harmonized Gurmukhi. Gurmukhi literally means "from the mouth of Guru" [17].

2.1. Structure of Gurmukhi Character

Gurmukhi script is written from top to bottom and left to right. Gurmukhi script is free from case sensitivity [13]. Gurmukhi words are partitioned into three zones; upper, middle and lower. The header line is responsible for separating the upper zone from the middle zone and the base line takes care of untying middle zone from the lower zone as shown in Fig. 1. The bottom strip comprises of lower modifiers [17].

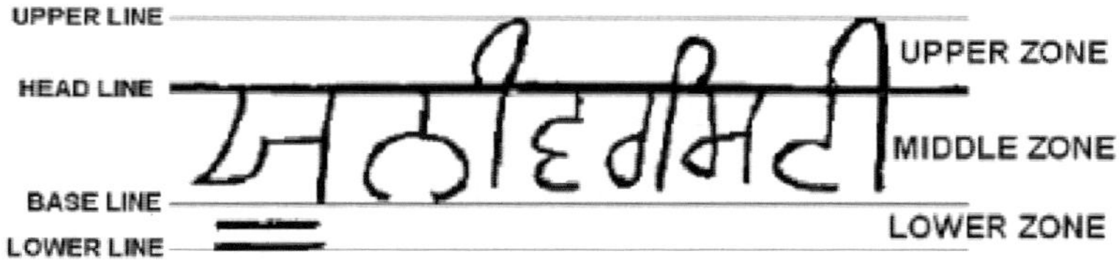

Figure 1. The Figure depicts three zones of a word in Gurmukhi script.

Gurmukhī has thirty-eight consonants (*akhar*), 10 vowel symbols (*lāgamātrā*), two symbols for nasal sounds (*pair bindi* and *ṭippī*), and one symbol which duplicates the sound of any consonant (*addak*) [15].

2.2. Machine written text (MWT) Vs. Handwritten text (HWT) in Gurmukhi

The differences between machine written text (MWT) and handwritten text (HWT) are summarized below [7], [8], [14].

- The number of strokes in machine written language is comparatively less than handwritten text.
- MWT preciseness makes it more readable as compared to HWT.
- The size and dimensions of HWT face variation where MWT does not.
- The available formats and fonts are much more in the case of MWT as compared to HWT.
- MWT does not face the spacing problem as in the case of HWT.

Fig. 2 depicts similar words written in MWT and HWT.

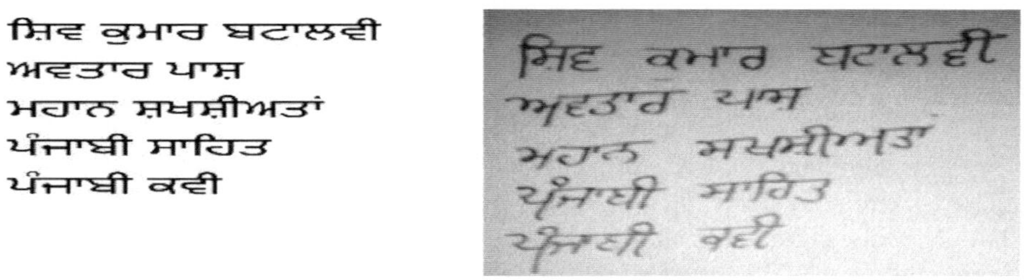

Figure 2. Displays MWT (left) against HWT (right)

3. Contribution and Implementation of research work

The motive of research work is to perform the comparison of HWT and MWT via designing appropriate program using simulation tool (MATLABR2013) on three different parameters; MSE (mean square error), PSNR (peak signal to Noise ratio), and IMV(intensity mean value).

(i) *SSIM (Structural Similarity Index Measure): SSIM* is responsible for calculating the similarity index between two images. The initial uncompressed image is used as a reference to decide the quality of images under study [4], [5].

IOP Conf. Series: Journal of Physics: Conf. Series **933** (2017) 012002 doi:10.1088/1742-6596/933/1/012002

(ii) *MSE (Mean Square Error):* It is a method used to check errors. The calculation is performed on two MSEs and compared to find the level of accuracy of an image. The quantitative score obtained helps to measure the degree of homogeneity or the level of error or distortion between them.

(iii) *PSNR(Peak Signal Noise Ratio):*PSNR is commonly used as a measure of the quality of reconstruction of loss compression codecs. The original data acts as a signal and the noise behaves as an error caused by compression. A higher PSNR would normally indicate that the reconstruction is of higher quality [4], [5].

3.1. Procedure
Primarily the image is scanned for preprocessing. Thereafter after removing noise from it, a suitable classifier rule is set up for HWT and MWT [9], [10]. The detailed steps are mentioned as under.

- Image is given as input from the external source.
- Image is converted into matrix comprised of rows and columns.
- RGB image is converted into gray scale image.
- Apply OTSU technique to identify threshold value of the concerned image.
- Labels of threshold image are identified.
- Perform extraction of text characters based on image labels.
- Segment text characters from the image.
- Apply three parameters (SSIM, PSNR, and MSE) for identifying MWT and HWT.
- Finally, the comparison is made.

3.2. *Implementation using MATLABR2013*
The detailed implementation of the research work is mentioned as under [2].Firstly, an imageis scanned for providing input to the system. The scanned image can be of HWT or MWT in Gurmukhi script [6].

Case 1: HWT

Let the input image be a handwritten image of the first letter of Gurmukhi script (*Oorra*) as shown in Fig. 3 [12].

Figure 3.Representation of first Gurmukhi HWC (*Oorra*)

Fig. 4 shows the screen of the developed program consisting two panels, left (HWT) and right (MWT). Each panel consists of three segments; Loading image, preprocessing and finally processing image. Fig. 4, shows the image been uploaded on the designed program on MATLABR2013 on HWT panel**.**

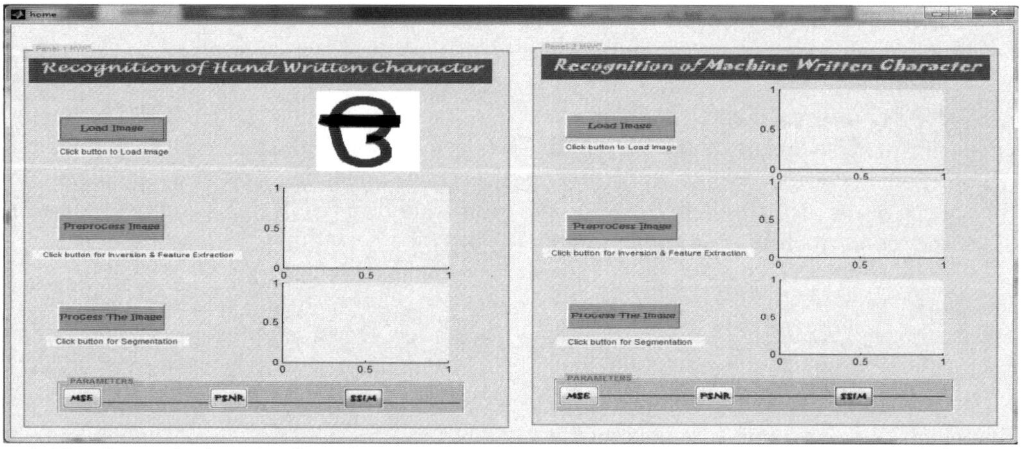

Figure 4. The figure depicts the first character (*Oorra*) of Gurmukhi script been uploaded on HWT panel.

After uploading the image, inversion of an image is performed followed by indexing and boxing as shown in Fig. 5.

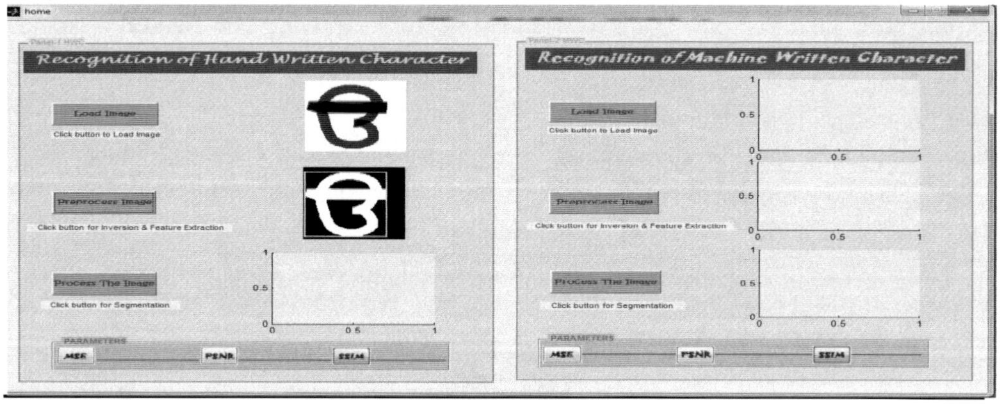

Figure 5.The figure depicts the image obtained after preprocessing in HWT panel.

Thereafter the image is processed which involves cropping, reshaping and resizing of the concerned image as shown in Fig. 6.

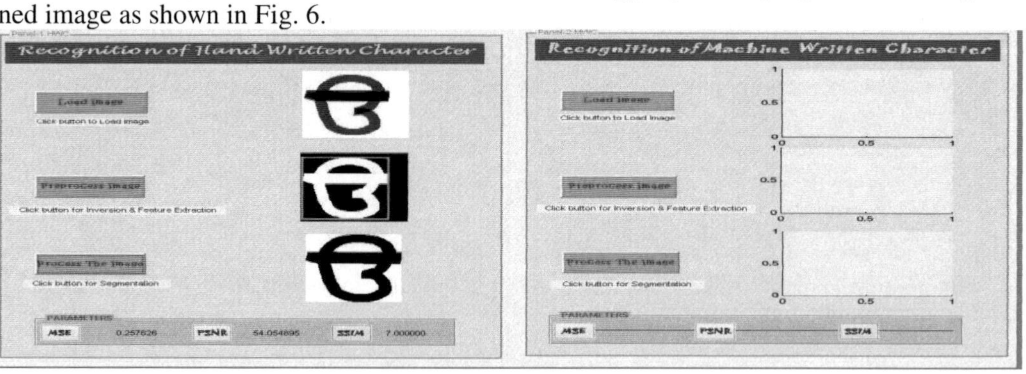

Figure 6. The figure displays the final processed image along with calculated parameters on HWT panel.

Finally, values are obtained for three different parameters under study as shown in Fig. 7.

Figure 7.The figure highlights the three different parameters values obtained on HWT panel.

3.3. *Case 2: MWT*

Firstly, provide input image to the system consisting machine written character (Fig. 8).

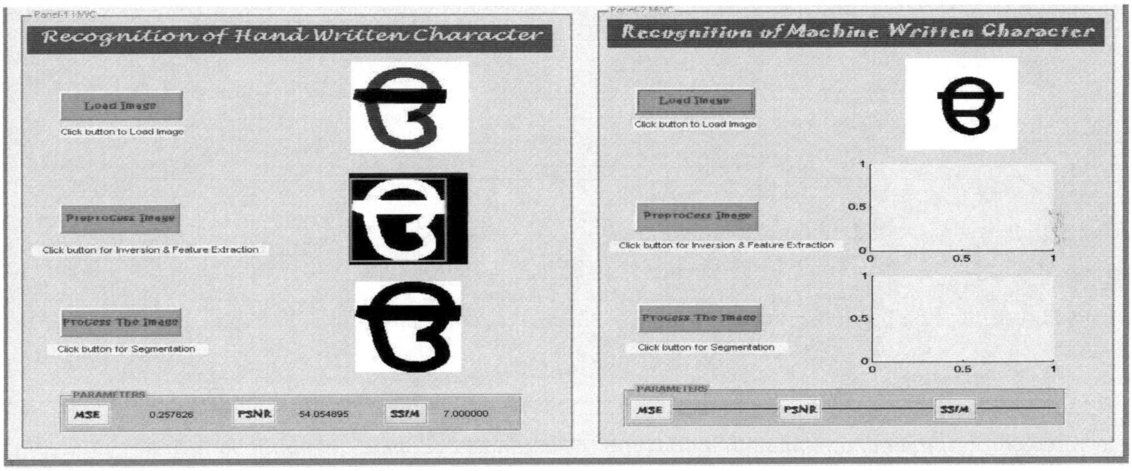

Figure 8.The figure depicts the first character (*Oorra*) of Gurmukhi script been uploaded on MWT panel.

Thereafter preprocessing of an image is conducted which comprises indexing, boxing, and cropping as shown in Fig. 9.

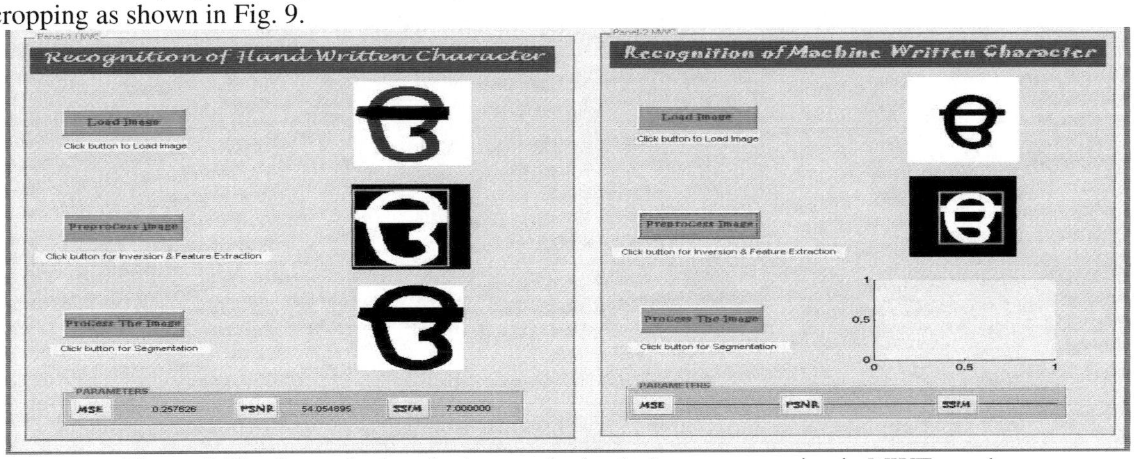

Figure 9.The figure depicts the image obtained after preprocessing in MWT panel.

Finally, the obtained image is cropped, reshaped and resized as shown in Fig. 10 followed by segmentation.

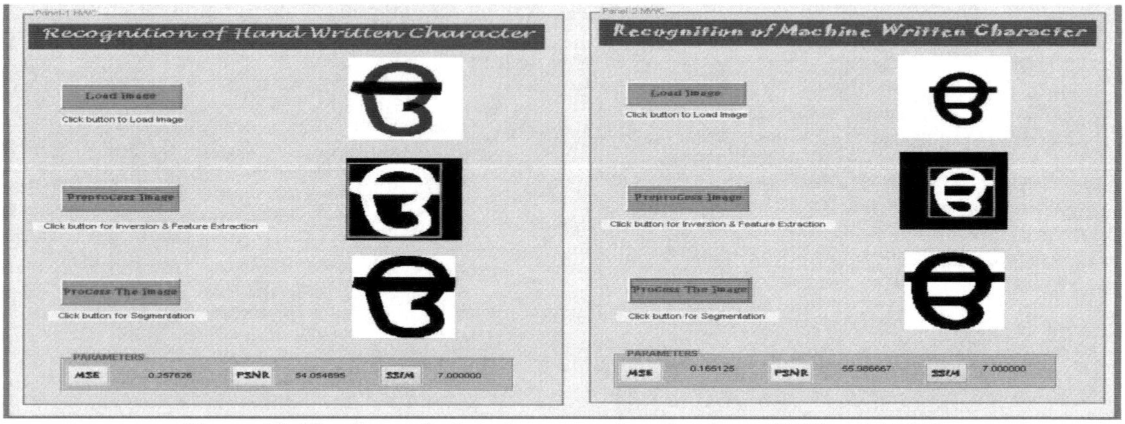

Figure 10. The figure depicts the comparison between HWT and MWT

Finally, values are obtained for three different parameters under study as shown in Fig. 11 in the case of MWT.

Figure 11. The figure highlights values of three different parameters obtained from MWT

3.4. *Comparing parameters – HWT Vs. MWT*

The values of the parameters obtained after running the designed code are compared in tabular form as shown in Fig. 12and in graphical form in Fig. 13.

	A	B	C
1		HWT	MWT
2	MSE	0.257626	0.165125
3	PSNR	54.054895	55.986667
4	SSIM	7	7

Figure 12. The figure shows the comparative values obtained from HWT and MWT based on parameters under study

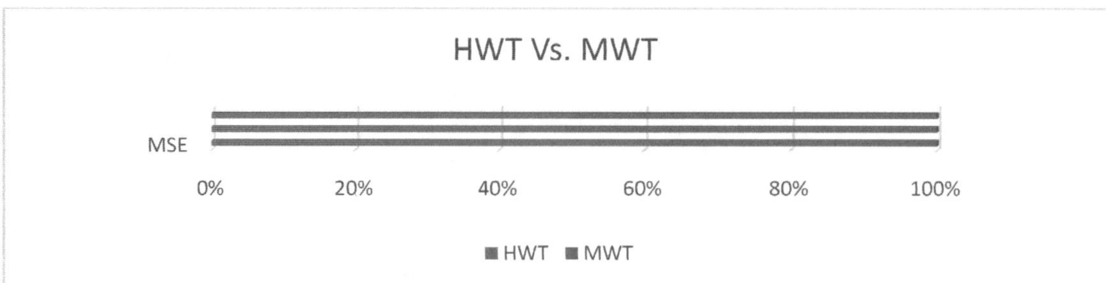

Figure 13. The figure shows the graphical comparison of HWT and MWT

As SSIM values face no change in value, the emphasis is laid on calculating MSE and PSNR values. Fig. 14 shows the MSE results of HWC and MWC obtained for next nine letters of Gurmukhi script followed by Fig. 15 in graphical form.

Gurmukhi Handwritten Character	MSE (HWC)	Gurmukhi Machine written Character	MSE (MWC)
ਅ	133.14286	ਅ	94.777778
ੲ	132.14286	ੲ	109.03175
ਸ	140.62698	ਸ	107.69048
ਹ	146.21429	ਹ	123.84127
ਕ	141.96032	ਕ	132.92857
ਖ	85.198413	ਖ	84.253968
ਗ	199.47619	ਗ	110.35714
ਘ	116.3254	ਘ	93.68254
ਙ	119.71429	ਙ	77.15873

Figure 14. MSE values of HWC and MWC for next nine letters in Gurmukhi script

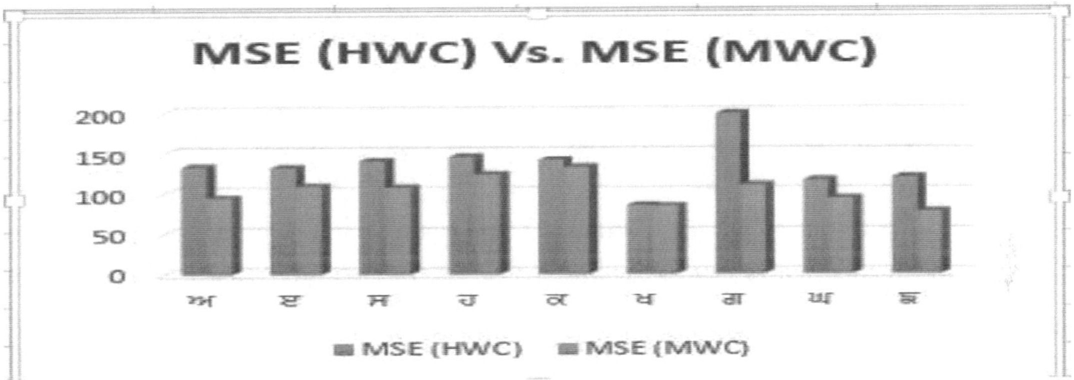

Figure 15. The figure displays a comparison of MSE values of HWT and MWT in graphical form.

Fig. 16, shows the PSNR results of HWC and MWC obtained for next nine letters of Gurmukhi script followed by Fig. 17 in graphical form.

Gurmukhi Handwritten Character	PSNR (HWC)	Gurmukhi Machine written Character	PSNR (MWC)
ਅ	26.921621	ਅ	28.397734
ੲ	26.931988	ੲ	27.78927
ਸ	26.684113	ਸ	27.843026
ਹ	26.514901	ਹ	27.236145
ਕ	26.64313	ਕ	26.928616
ਖ	28.860484	ਖ	28.908896
ਗ	25.165889	ਗ	27.736795
ਘ	27.508054	ਘ	28.448213
ਙ	27.38334	ਙ	29.290449

Figure 16. PSNR values of HWC and MWC for next nine letters in Gurmukhi script

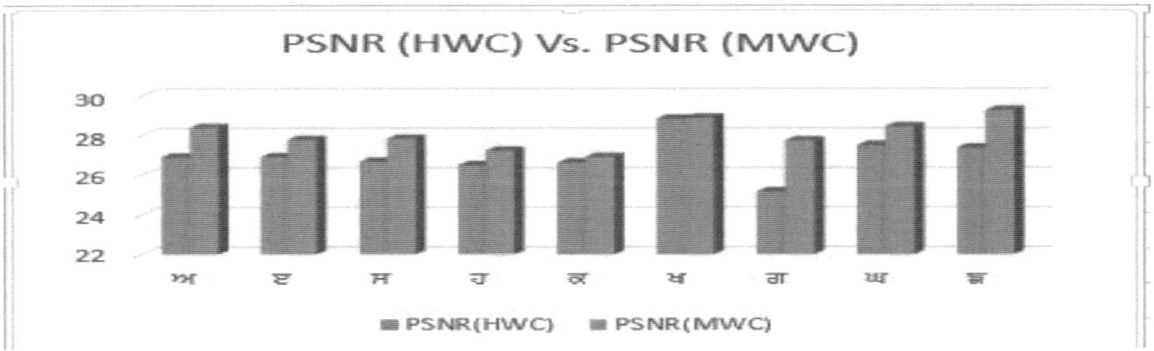

Figure 17. The figure displays a comparison of PSNR values of HWT and MWT in graphical form.

4. Conclusion

The results obtained from the research work performed is summarized as under.

- The MSE in the case of HWT is more than MWT.
- The value of PSNR is greater in the case of MWT as compared to HWT which indicates that reconstruction in MWT is of higher quality.
- SSIM value obtained in both the cases is same.

5. References

[1] Chaudhari, A., Krupa, M., Badelia, P., Soumya .K.G. (2017). *Optical Character Recognition Systems for Different Languages with Soft Computing*. Springer International Publishing.

[2] Jindal, A. & Amir, M. (2014). Automatic classification of handwritten and printed text in ICR boxes.*Proceedings of the IEEE International: Advance Computing Conference* (pp. 1028 - 1032).

[3] Optical Character Recognition (OCR). http://www.explainthatstuff.com/how-ocr-works.html. Date accessed: 06/06/2017.

[4] Vij, K.,Singh, Y. Enhancement of Images using Histogram processing techniques. *International Journal of Computer Technology and Applications,* 2 (2), 309 - 313.

[5] Jin, S., Gaoding, N. (2012). *Signal processing using the wavelet transform and the Karhunen-Loeve transform.* Kristianstad University, Sweden.

[6] Rani, M., Meena, Y.K. (2011). An efficient feature extraction method for handwritten character recognition. *Proceeding of the second International Conference on Swarm, Evolutionary and Memetic Computing: Vol. Part II.* (pp. 302 - 309).

[7] Kumar, M., Jindal, M.K., Sharma, R.K. (2012). Weka based Offline Handwritten Gurmukhi Character Recognition. *Proceedings of International Conference on Soft Computing for Problem* Solving (pp. 711 - 722).

[8] Kumar, M., Sharma, R.K., Jindal, M.K. (2011). SVM based Offline Handwritten Gurmukhi Character Recognition. *Proceedings of International Workshop on Soft Computing Applications and Knowledge Discovery* (pp. 51 - 62).

[9] Narayan, S., Nikolaidis, A. (2012). Discrimination of handwritten and machine Printed text is Scanner document Images based on Rough Set Theory. *Proceedings of the IEEE: 10th workshop on Multimedia Signal processing* (pp. 393 - 398).

[10] Pal, U., &Chaudari, B.B. (1999). Machine-printed and hand-written text lines identification. *Proceedings of IEEE: Fifth Int. Conference on Document Analysis and Recognition* (pp. 645 - 648).

[11] Saba, T., &Nikolaidis, A. (2015). Language Independent Rule Based Classification of Printed & Handwritten Text. *Proceedings of the IEEE: Tenth workshop on Multimedia Signal Processing* (pp. 393 - 398).

[12] Li, Z., & Tan, C.L. (2009). Character recognition under severe perspective Distortion. *Proceedings of 10th International Conference on Document Analysis and Recognition* (pp. 676 - 680).

[13] Kaur, P., & Singh, B. (2012). Recognition of signboard images of Gurmukhi. *Journal of Global Research of Computer Science*, 20-23.

[14] Kumar, M., Jindal, M. K., & Sharma, R. K. (2014). A Novel Hierarchical Techniques for Offline Handwritten Gurmukhi Character Recognition**.** *National Academy Science Letters*, 567 - 572.

[15] Kumar, N., & Gupta, S. (2016). Offline Handwritten Gurmukhi Character Recognition: A Review. *International Journal of Software Engineering and Its Applications,* 77-86.

[16] Garg, N. (2009). *Handwritten Gurumukhi Character Recognition Using Neural Networks.* Postgraduate dissertation, Thapar University, Patiala.

[17] Importance of handwritten documents and machine written documents. (2017, March 28). *Daily Punjab Times*, pp. 4.

IOP Conf. Series: Journal of Physics: Conf. Series **933** (2017) 012003 doi:10.1088/1742-6596/933/1/012003

The role of the coarsest resolution subband in the wavelet-based reconstruction of signals from gradients

Ioana S. Sevcenco and Panajotis Agathoklis

Electrical and Computer Engineering Department, University of Victoria, Canada

ioana.sevcenco@ece.uvic.ca

Abstract. An efficient wavelet-based algorithm to reconstruct non-square/ non-cubic signals from gradient data is proposed. This algorithm is motivated by applications such as image or video processing in the gradient domain. In some earlier approaches, the non-square/ non-cubic gradients were extended to enable a square/ cubic Haar wavelet decomposition and the coarsest resolution subband was derived from the mean value of the signal. In this paper, a non-square/non-cubic wavelet decomposition is obtained directly without extending the gradient data. The challenge comes from finding the coarsest resolution subband of the wavelet decomposition and an algorithm to compute this is proposed. The performance of the algorithm is evaluated in terms of accuracy and computation time, and is shown to outperform the considered earlier approaches in a number of cases. Further, a closer look on the role of the coarsest resolution subband coefficients reveals a trade-off between errors in reconstruction and visual quality which has interesting implications in image and video processing applications.

1. Introduction

In many interesting applications from adaptive optics [1], image processing [2], [3], [4] or video processing[5], [6] signal reconstruction from gradient data plays an important role. In some of these applications, such as adaptive optics, the data available is in the form of gradient values instead of signal values. In others, dealing with the human visual system, inherently highly sensitive to contrast, the gradient domain offers an attractive solution domain. Working in the gradient domain, however, requires reconstructing a signal from gradient data. Signal reconstruction from gradient data has been reformulated as a Poisson equation in [2], and several methods have been developed for signal reconstruction. A class of Fourier based techniques[7] with typical complexity $O(N\log N)$, where N is the number of unknowns, have been presented. An algebraic approach based on graph theory is proposed in [8]. Another class consists of iterative solvers [9], which include methods such as Jacobi and Gauss-Seidel, but these are typically slow to converge on large scale problems. A very popular class of fast solvers is the multigrid approach [10], which solves the Poisson problem on diagonally oriented grids, and uses an iterative Poisson solver to smooth the error at each scale and perform the interpolation to obtain solution estimates on the finest grid. Multigrid techniques are $O(N)$.

A wavelet based reconstruction technique was proposed in [11], and used in the context of adaptive optics. This approach also has linear complexity, and was developed and adapted forimage [12] and video [5] processing applications. This method is based on obtaining the Haar wavelet decomposition directly from the gradient data. The signal can then be obtained from this wavelet decomposition. This method deals with non-square (in 2-D) or non-cube (in 3-D) signal by expanding the non-square signal to a square (or cube) of appropriate size. Although this expansion yields satisfactory results, it may require large amounts of additional memory, particularly when editing long video sequences. In this paper, this wavelet-based reconstruction approach is expanded to deal with non-square and non-cube cases without expanding the given data. The approach is based on obtaining the coarsest resolution subband (low-low) of the wavelet decomposition directly from the gradient data. This new approach gives also the opportunity to study closer the effect of errors in the low-low part of the wavelet decomposition. The effects of such errors will be illustrated with examples and the insight gained provides important recommendations for more efficient ways of reconstructing non-square/ non-cubic signals from gradient data, with applications in image or video processing.

The remaining of this paper is organized as follows: Section 2 briefly reviews the wavelet based reconstruction technique. In Section 3, a method is proposed to deal with non-square/non-cube signals. In Section 4, the performance of the technique is evaluated and discussed with respect to accuracy, visual quality and computational resources required. Conclusions are drawn in Section 5.

2. Background information

The Haar wavelet analysis filters are: $H_L(z)=\left(1+z^{-1}\right)/\sqrt{2}$ and $H_H(z)=\left(1-z^{-1}\right)/\sqrt{2}$. The reconstruction technique in [11], [12], [5]has an analysis and a synthesis step. In the analysis step, the Haar wavelet decomposition of the signal is obtained from the gradient data and the average value of the signal. This step is based on the relation between the gradient and the Haar wavelet analysis filters:

$$\Phi_d=\sqrt{2}\Phi\otimes H_H\left(z_d\right) \tag{1}$$

where Φ_y, Φ_x are gradient components, and $\Phi\otimes H_H\left(z_d\right)$ denotes 1-D convolution of 2-D signal Φ with high pass analysis filter $H_H(z)$ along direction $d\in\{y, x\}$. This observation remains valid in higher dimensions. In the synthesis step, the signal is recovered from the wavelet decomposition by wavelet synthesis, with the possibility of including an iterative Poisson solver (Jacobi [9]) at each resolution. Including the iterative Poisson solver is important in gradient based image or video editing applications, where gradient data is typically altered and the zero-curl condition is not satisfied.

In the analysis step of [12] the gradient of non-square signals is first extended to square and then the Haar wavelet decomposition is obtained from the extended data. This has non-negligible implications on memory consumption and computation time. While extending to square is still an option in 2-D, the data size quickly becomes a problem for 3-D signals with significant differences in size, such as video sequences. The reconstruction algorithm proposed in Sec. 3 does not entail signal extension, making this approach more suitable for signals with non-negligible size differences.

3. Obtaining the coarsest resolution subband coefficients from signal derivatives

The first step of the reconstruction is to obtain the Haar wavelet decomposition of the signal directly from the derivatives. For a non-square/ non-cubic signal with size $2^M\times 2^N$ (or $2^M\times 2^N\times 2^P$) the equations for finding the "low-high", "high-low" and "high-high" subbands at all resolutions of the Haar wavelet decomposition from the given gradient components are the same as those for a square (or cubic) signal in[12], [5]provided that the maximum number of levels in the decomposition is M (if $M<N$). What is different in the non-square (non-cubic) case is the way in which the "low-low" (or low-low-low) subband at the coarsest resolution of the Haar wavelet decomposition is found. These subband coefficients are referred to as the coarsest resolution subband. For square (or cubic) signals, the coarsest resolution subband is a scalar proportional to the mean value of the signal. For non-square (or non-cubic) signals, the coarsest resolution subband is a multi-dimensional array.Finding this array of numbers from the signal derivatives and the mean value of the signal is presented next.

3.1. The two dimensional case

Let Φ be a 2-D signal, with size $2^M \times 2^N$,where $1 < M \le N$ are integers. Let y and x be two orthogonal directions. The coarsest resolution low-low subband of the Haar wavelet decomposition of Φ, denoted $\Phi^M_{L_y L_x}$, is a signal with size $1 \times 2^{N-M}$, obtained by successive filtering with $H_L(z)$ and subsampling. It can be shown that it is proportional to the partial sums of all elements in consecutive regions of Φ, as illustrated in Fig. 1. The objective of this work is to find $\Phi^M_{L_y L_x}$ from the signal derivatives and its mean value.

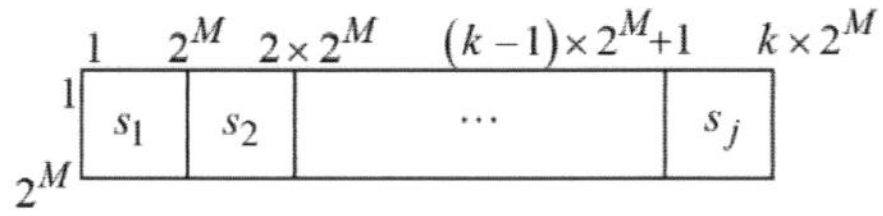

Fig. 1. Representation of $\Phi^M_{L_y L_x} = \left[s_1, s_2, \ldots, s_j\right]$ for rectangular signals with size $2^M \times 2^N$ ($M \le N$), $j = N - M$ and $k = 2^j$.

Let the given first order directional derivatives of the unknown signal be denoted by Φ_y and Φ_x (forward difference approximation). Let m be the mean value of the unknown signal Φ. Finding $\Phi^M_{L_y L_x}$ amounts to finding the $N - M$ partial sums of signal regions illustrated in Fig. 1. This leads to the following algorithm:

Step 1. Compute:

$$c_x = A_x \Phi_x^T B_x \tag{2}$$

where $A_x = \begin{pmatrix} 1 & 2 & 1 & 0 & 0 & \cdots & 0 & 0 & 0 \\ 0 & 0 & 1 & 2 & 1 & \cdots & 0 & 0 & 0 \\ \vdots & \vdots & \vdots & \vdots & \vdots & \ddots & \vdots & \vdots & \vdots \\ 0 & 0 & 0 & 0 & 0 & \cdots & 1 & 2 & 1 \end{pmatrix} \in \mathbb{R}^{\left(2^{N-1}-1\right)\times\left(2^N-1\right)}$, and $B_x = \begin{bmatrix} 1 & 1 & \cdots & 1 \end{bmatrix}^T \in \mathbb{R}^{2^M \times 1}$.

Step 2. Compute:

$$u = \frac{1}{2^M} A^{-1} v \tag{3}$$

where: $A = \begin{bmatrix} 1 & 1 & 1 & 1 & \cdots & 1 & 1 & 1 \\ -1 & 1 & 0 & 0 & \cdots & 0 & 0 & 0 \\ 0 & -1 & 1 & 0 & \cdots & 0 & 0 & 0 \\ 0 & 0 & -1 & 1 & \cdots & 0 & 0 & 0 \\ \vdots & \vdots & \vdots & \vdots & \ddots & \vdots & \vdots & \vdots \\ 0 & 0 & 0 & 0 & \cdots & -1 & 1 & 0 \\ 0 & 0 & 0 & 0 & \cdots & 0 & -1 & 1 \end{bmatrix} \in \mathbb{R}^{2^{N-1} \times 2^{N-1}}$ and $v = \begin{bmatrix} m \cdot 2^{M+N} \\ c_x \end{bmatrix} \in \mathbb{R}^{2^{N-1} \times 1}$.

Step 3. Compute:

$$\Phi^M_{L_y L_x} = u^T B \tag{4}$$

where $B = \begin{bmatrix} \mathbf{B}_1 & \mathbf{B}_2 & \cdots & \mathbf{B}_k \end{bmatrix}^T$, $k = 2^{N-M}$ and $\mathbf{B}_p = \begin{bmatrix} 0 & \cdots & 0 & 1 & 0 & \cdots & 0 \\ \vdots & \ddots & \vdots & \vdots & \vdots & \ddots & \vdots \\ 0 & \cdots & 0 & 1 & 0 & \cdots & 0 \end{bmatrix} \in 2^{M-1} \times 2^{N-M}$, in which the pth column is an all-ones vector with $M - 1$ elements.

3.2. The three dimensional case

Let Φ be a 3-D signal, with size $2^M \times 2^N \times 2^P$ where $1 < M \leq N \leq P$ integers. Signal value at a point (y, x, t) is denoted $\Phi(y, x, t)$ with *y*, *x* and *t* orthogonal directions. The coarsest resolution subband of the Haar wavelet decomposition of Φ is denoted $\Phi^M_{L_y L_x L_t}$ and is a signal with size $1 \times 2^{N-M} \times 2^{P-M}$. This signal can be found similarly to the 2-D case, by solving a system of linear equations. This system of equations is obtained from the given discrete partial derivatives and the mean value of the signal. More precisely, the partial derivative Φ_x can be used to derive $2^{P-M}\left(2^{N-1}-1\right)$ equations. and the partial derivative Φ_t can be used to derive $2^{N-M}\left(2^{P-1}-1\right)$ equations. The mean value of the signal yields one more equation. In this way, a system of linear equations is formed:

$$A\boldsymbol{\lambda} = B \tag{5}$$

where: $A \in \mathbb{R}^{\left[2^{P-M}\left(2^{N-1}-1\right)+2^{N-M}\left(2^{P-1}-1\right)+1\right]\times 2^{N+P-2M}}$ and $B \in \mathbb{R}^{\left[2^{P-M}\left(2^{N-1}-1\right)+2^{N-M}\left(2^{P-1}-1\right)+1\right]\times 1}$ are known quantities obtained from partial derivatives and the mean value of the signal, and $\boldsymbol{\lambda}$ is the unknown, which, once found is reshaped as $\Phi^M_{L_y L_x L_t}$. Due to space limits details are not included.

4. Performance evaluation

This section analyzes the performance of different reconstruction algorithms and the role of the coarsest resolution subband coefficients on the quality of signals reconstructed from gradient data. The comparison is between: Algorithm 1: the method in Section 3.1; Algorithm 2: the method in [12], [5]where non-square/ non-cubic gradient data is extended to nearest square/ cube; Algorithm 3: the signal is reconstructed using a coarsest resolution subband with all entries equal to a value proportional to the signal mean. Algorithm 3 is particularly interesting in applications such as image or video editing, as it avoids both the use of additional memory required by data extension to square/ cube and the computations for theexact coarsest resolution subband. Results show that although this approach leads to lower accuracy reconstructed signals, they are still visually acceptable, if an iterative Poisson solver is included in the synthesis step. The benefits of Algorithm 3 become significant in the case of 3-D data such as video sequences, as will be seen shortly.

The reconstruction quality will be determined by analyzing the signal to noise ratio (SNR), measured in dB, between the original signal and the estimate obtained from the derivatives and the mean value of the signal, as an estimate of solution accuracy; and the time required for a MATLAB implementation of the methods, as an estimate of computational speed.

4.1. The two dimensional case

Reconstruction accuracy was first considered for Algorithms 1 and 3, and the results are shown in Fig. 2. The reconstruction accuracy of Algorithm 2 is similar to that of Algorithm 1, and the results were omitted in the interest of not cluttering the figure. In Algorithm 3, the values of $\Phi^M_{L_y L_x}$ are left 0 before wavelet synthesis, and the average value of the reconstruction is corrected at the end, to match the given average. This corresponds to the partial sums $s_1, \ldots s_j$ in Fig. 1 all having the same value. As can be seen from Fig. 2, obtaining the values of $\Phi^M_{L_y L_x}$ exactly from the input data results in a very high SNR in the reconstructed signal, and, as expected, leaving the coarsest resolution "low-low" subband coefficients zero before synthesis and adjusting the signal average value at the end significantly lowers the SNR of the reconstructed signal.

Run times of MATLAB implementations of the three reconstruction algorithms were compared. As Fig. 3 reveals, the performance is similar for square signals, as signal extension is not performed in this case. The advantage of the new approach (i.e., Algorithms 1 and 3) becomes clear when dealing with non-square signals, as a solution is produced faster than when Algorithm 2 is used.

Let us now take a closer look at the reconstruction error in Algorithm 3. In Fig. 4a,a 64×256 image was obtained from image "threads"[13]. The image reconstructed from its gradient and mean value using Algorithm 1 is shown in Fig. 4b, and is an accurate representation of the image in Fig. 4a. The image reconstructed using Algorithm 3 is shown in Fig. 4c and has visual artifacts due to not obtaining exact values of the coarsest resolution "low-low" subband coefficients. The image reconstructed using Algorithm 3, with an iterative Poisson solver included at each resolution in the synthesis step is shown in Fig. 4d. Including the Poisson solver has the effect of smoothing out the vertical lines noticeable in Fig. 4c. The objective quality of each image was also evaluated by comparison with the original image, using SNR and SSIM[14]. The SNR values agree with the noticeable increase in visual quality (from Fig. 4c to 4d). It is interesting to note that SSIM ranks the result in Fig. 4d of lower quality than that in Fig. 4c, although visual examination indicates otherwise.

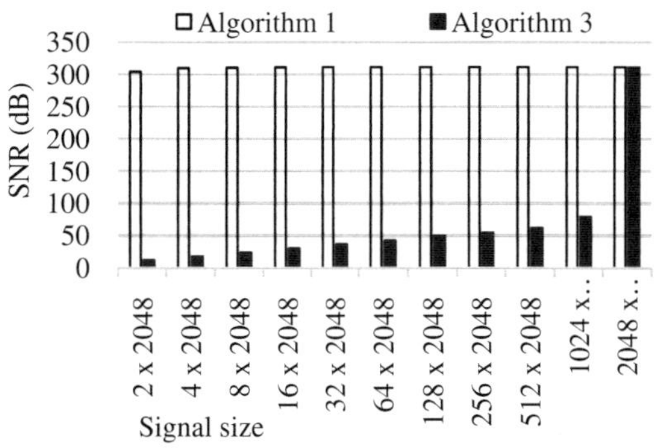

Fig. 2. Performance evaluation in terms of solution accuracy.

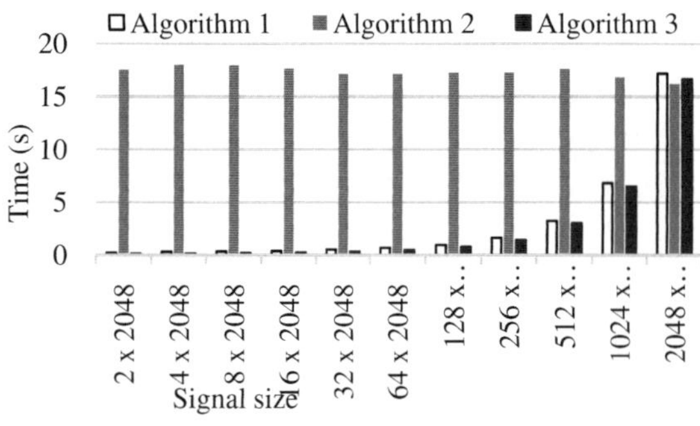

Fig. 3. Performance evaluation in terms of speed.

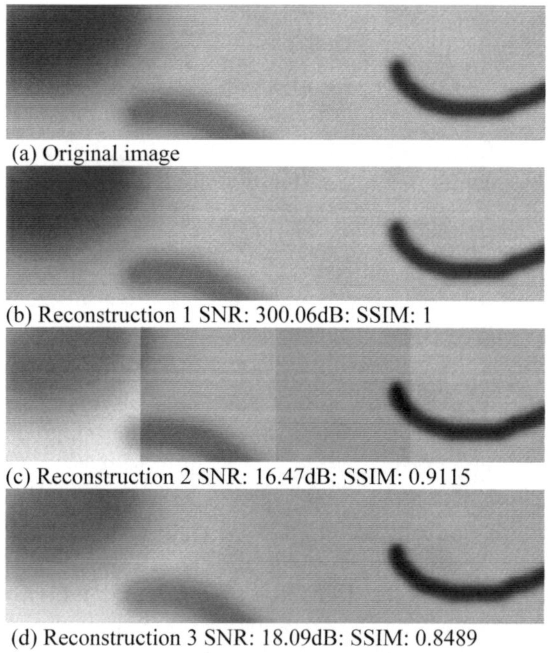

(a) Original image

(b) Reconstruction 1 SNR: 300.06dB: SSIM: 1

(c) Reconstruction 2 SNR: 16.47dB: SSIM: 0.9115

(d) Reconstruction 3 SNR: 18.09dB: SSIM: 0.8489

Fig.4. Visualizing the role of the coarsest resolution low subband coefficients and of the Poisson solver. Top to bottom: (a) original image; (b) image reconstructed with exact calculation of $\Phi^{M}_{L_yL_x}$; (c) image reconstructed without exact calculation of $\Phi^{M}_{L_yL_x}$ without Poisson solver; (d) image reconstructed without exact calculation of $\Phi^{M}_{L_yL_x}$, with Poisson solver in the wavelet synthesis.

These experiments reveal the main trade-off of the new approach, and a recommendation can be made. If signal accuracy is of primary importance for the application at hand, the coarsest resolution subband coefficients should be computed exactly before wavelet synthesis. If memory requirements and speed are more important, the coarsest resolution subband coefficients can be approximated, and the result is still visually acceptable if the Poisson solver is included in the wavelet synthesis step.

4.2. The three dimensional case

Reconstructions obtained by computing the coarsest resolution subband using the method in Sec. 3.2. (denoted "With LLL" in Fig. 5) are compared with reconstructions obtained using the coarsest resolution subband with all entries equal to a value proportional to the signal mean (denoted "Without LLL" in Fig. 5). The Poisson solver is included at each resolution. In terms of solution accuracy (SNR) the results are similar to the ones of the 2-D case. In terms of speed, the reconstructions obtined "Without LLL" are compared to the earlier technique in [5], with signal extension to nearest cube. The results are shown in Fig. 6. The results of Fig. 5, 6 indicate that for 3-D signals with large differences between the three dimensions, the new approach produces a solution in significantly less time than the approach in [5]. This is expected since a significant amount of memory is needed to store the extended version of the non-cubic signals, and this requires additional processing time.

5. Conclusions

An efficient way of reconstructing non-square/ non-cubic signals from gradient data was presented. The approach is based on finding the Haar wavelet decomposition of the signal first, and then reconstructing the signal by wavelet synthesis. The focus is on obtaining the coarsest resolution subband of the wavelet decomposition, and on studying the role of these coefficients on the quality of the reconstructions. The new algorithm was compared with a previous reconstruction approach, which extends gradient data to the nearest square or cube. The experiments show that obtaining an exact approximation of the coarsest resolution coefficients generally increases solution accuracy to the expense of computation time. However, studies reveal that in applications such as image or video editing, finding these coefficients exactly has no significant consequence on the visual quality of the reconstructed signal. This finding has great potential in higher dimensional signal processing applications such as light fields, and this is the next step in our research.

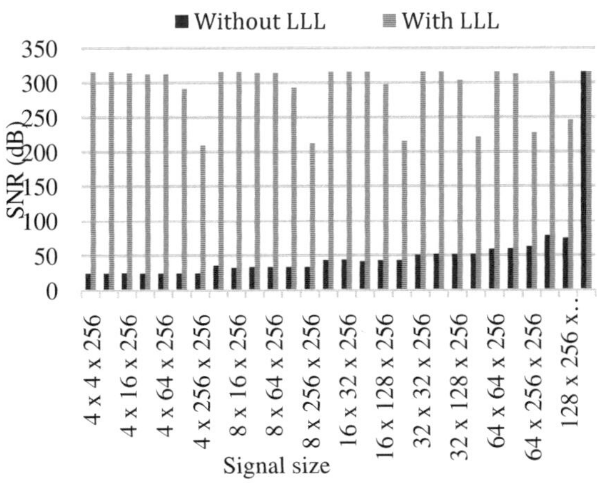

Fig. 5. Performance evaluation in terms of solution accuracy. 3-D case. Poisson solver used in the synthesis.

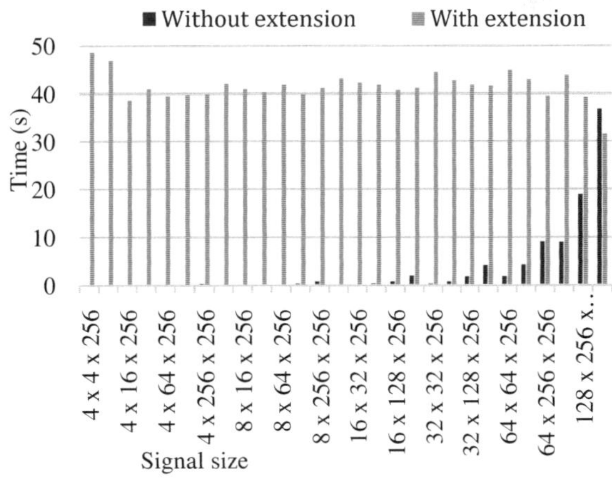

Fig. 6. Performance evaluation in terms of speed. 3-D case. Poisson solver used in the synthesis.

6. References

IOP Conf. Series: Journal of Physics: Conf. Series **933** (2017) 012003 doi:10.1088/1742-6596/933/1/012003

[1] Redding S and Bikkannavar D 2010 The end of the blur *IEEE Spectrum* **47** pp 46–53

[2] Fattal R, Lischinski D and Werman, M 2002 Gradient domain high dynamic range compression *Proceedings of the 29th Annual Conference on Computer Graphics and Interactive Techniques SIGGRAPH '02* pp 249–256

[3] Pérez P, Gangnet M and Blake A 2003 Poisson image editing *ACM Trans. Graph.* **22** pp 313–318

[4] Levin A, Zomet A, Peleg S and Weiss Y2004 Seamless image stitching in the gradient domain *Proceedings of European Conference on Computer Vision* pp 377–389

[5] Sevcenco I and Agathoklis P 2015 Video editing in the gradient domain using a wavelet based 3-D reconstruction algorithm and an iterative Poisson solver *IEEE Pacific Rim Conference on Communications, Computers and Signal Processing* pp 205–209

[6] Wang H, Xu N, Raskar R and Ahuja N2005 Videoshop: a new framework for spatio-temporal video editing in gradient domain *Graph. Models***69** pp 57–70

[7] Frankot R and Chellappa R, 1988 A method for enforcing integrability in shape from shading algorithms *IEEE Transactions on Pattern Analysis and Machine Intelligence* **10** pp 439–451

[8] Agrawal A, Chellappa R and Raskar R2005 An algebraic approach to surface reconstruction from gradient fields *Tenth IEEE International Conference on Computer Vision***1** pp 174–181

[9] Saad Y2003 Iterative Methods for Sparse Linear Systems Philadelphia SIAM

[10] Roberts A2001 Simple and fast multigrid solution of Poisson's equation using diagonally oriented grids *ANZIAM J* **43**pp 1–36

[11] Hampton P, Agathoklis P and Bradley C2008 A new wave-front reconstruction method for adaptive optics system using wavelets *IEEE Journal of Selected Topics in Signal Processing* **2**781–792

[12] Sevcenco I, Hampton P and Agathoklis P, 2015 A wavelet based method for image reconstruction from gradient data with applications *Multidimensional Systems and Signal Processing* **26** 717–737

[13] Mathworks 2017 Image Processing toolbox of MATLAB R2017a.

[14] Wang Z, Bovik AC, Sheikh H R and Simoncelli E P 2004 Image quality assessment: from error visibility to structural similarity *IEEE Transactions on Image Processing* **13**pp 600–612

Acknowledgments
This work was in part supported by the Natural Sciences and Engineering Research Council of Canada (NSERC), and by Mr. John Montalbano, through a donor award to the University of Victoria.

IOP Conf. Series: Journal of Physics: Conf. Series **933** (2017) 012004 doi:10.1088/1742-6596/933/1/012004

Segmenting Bone Parts for Bone Age Assessment using Point Distribution Model and Contour Modelling

AmandeepKaur [a] and Dr. Kulwinder Singh Mann [b]

[a]Research Scholar (CSE), I.K. Gujral Punjab Technical University, Jalandhar, India; [b]Professor and Dean, Department of Information Technology, Guru Nanak Dev Engineering College, Ludhiana, India.

Email: [a]rs.amandeepkaur@ptu.ac.in, [b]mannkulwinder@gndec.ac.in

Abstract. Bone age assessment (BAA) is a task performed on radiographs by the pediatricians in hospitals to predict the final adult height, to diagnose growth disorders by monitoring skeletal development. For building an automatic bone age assessment system the step in routine is to do image pre-processing of the bone X-rays so that features row can be constructed. In this research paper, an enhanced point distribution algorithm using contours has been implemented for segmenting bone parts as per well-established procedure of bone age assessment that would be helpful in building feature row and later on; it would be helpful in construction of automatic bone age assessment system. Implementation of the segmentation algorithm shows high degree of accuracy in terms of recall and precision in segmenting bone parts from left hand X-Rays.

1. Introduction

To solve complex real world problems and challenges, the scientific communities focus on building specialized machine learning algorithms but these algorithms can work effectively only if appropriate datasets are first created and pre-processed to suit the machine learning algorithms. These features are normally based on evaluation of single or multiple predictor variables. In case of bone age assessment the predictors normally are the dimensions, size and gaps between the different bones of specific part of skeleton such as wrist and hands. By using bone age charts and atlases [1], data analysis is often performed in a highly manual and subjective manner using very limited automatic image analysis and it depends more on the skill and experience of the bone age assessor and the process also suffers from inter-rater variability problems and other biases that affect the calibrations of bones' dimensions. Hence, there is the need of automatic bone age assessment systems [2][3][4].

2. Related Work

According to the recent citations, it has been observed that the computation on human bones to access their maturity itself is a challenge. This can be attributed to many factors such as ossification and fusion of different hand bones at different stages but the main determinants are their races, diet habits, environment etc. Hence, region wise bone age charts or atlases should be built [1]. Technically, the image processing functions such as noise removal, background subtraction, soft tissue identification have undergone many changes to finally arrive at the set of measurement for accessing bone age[6]. Evaluation of the threshold based methods for segmentation has not resulted in fair degree of accuracy but adaptive thresholding methods may provide better results. Many algorithms are unable to find

IOP Conf. Series: Journal of Physics: Conf. Series **933** (2017) 012004 doi:10.1088/1742-6596/933/1/012004

gradient difference in the soft tissue[7] for segmenting the bone parts. The region growing and merging methods are the methods that are dependent on the similarity[8] metric and connectivity[9] that carries a risk of fusing the epiphysis sites (situated around the metaphysic [2]).Secondly, region-based segmentation is insensitive to the semantics of the bone parts in the image. In simple words, it means that it does not recognize the membership based on pre-defined membership functions (defining the bone part). Each bone part, if considered as homogenous region[5] does not work well as per assumption that bone is formed by cancellous bone and cortical bone that has high variations on texture and intensity range. Some authors have worked on segmenting carpal bones [9] and deformable models such as active contour, active shape modeling[10][11][12] and active appearance modeling. In bio-medical application most the shapes of the organs, bones parts are not simple morphological boundaries and but have complex or arbitrary geometries. For, such it is always desirable to build a training dataset using a medical expect. Thus, in this research, these well established, proven methods are proposed to be used for bone part(s) segmentation as it will allow building of specific shape & approach models for each bone part using medical expertise.

3. Designing of algorithm for segmenting bone parts

In this section, the method employed for determining the edges of different bone structure/part has been discussed [figure 1]. There is a need to build a separate Point Distribution Model to segment each bone part[14][15] that can help to identify the contours[16][10] or boundaries of the each object so that it can finally be segmented into separate frames for further feature analysis. It should also be noted that all images undergo a preprocessing steps to make them standardized and workable. The x-ray image of left hand has been adjusted into a square of 800x600 consisting of full left hand image and 240x240 pixel of particular anatomical region.

3.1. The formulation of the local edge boundary: Landmarks or anchor points are marked on the image that represent the boundary of bone part in the following manner :

a) Let 'lmp.n' number of major land mark points 'lmp' manually clicked.
b) Let (lmp.x, lmp.y) be the location of the landmark points clicked.
c) Let Imp.I be variable representing the Image of Bone Part.
d) Interpolate to get more pairs of points (lmp.x,lmp.y).
e) Mark the Landmark points with 1, other points as zero.
f) Interpolate new set of landmark points and evenly space them.

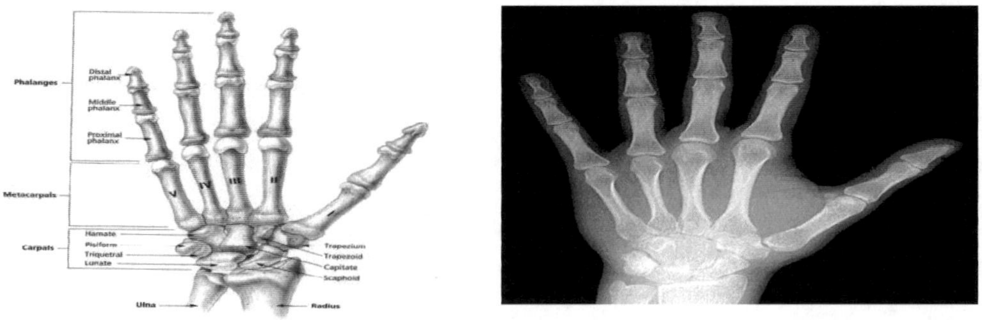

Figure 1(a). Anatomical Locations of the bones parts in hand,(b) X-Ray of 18 Year Old Subject

IOP Conf. Series: Journal of Physics: Conf. Series **933** (2017) 012004 doi:10.1088/1742-6596/933/1/012004

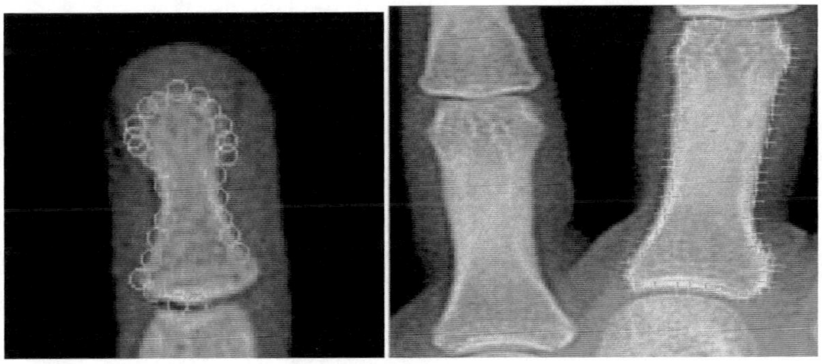

Figure2: Major Landmarks of Distal Phalanges and Meta Carpals bones

This process is a method of understanding the distribution of data points in a vector space model. These landmark points[15] are the most prominent points from which the shape of the bone part can be drawn given a locus. These points (x, y) become instances of the training set.

This collection of landmark points (anchor points) is the mean geometry of the shape of the bone part [Figure 2] with statistical model. It has been empirically found that the shape of bone instances varies statistically but the key features of the bone shape remain more or less same. Therefore, the constitute of bone shape is its data points represented using mathematical functions. This step gives the landmarks data points and the lines representing dataset which is divided into training and validation datasets.

3.2. Formulating the contour shape based model: The distribution of data points forms a contours or a distinct boundary [Figure 2]. In this step a new image is constructed by matching the statistical model of the bone part and its appearance model. The images with multiple scales are taken and the model is built from smaller scale to largest scale. Due to this multiple-resolutions approach variations in the data need to be smoothen and the construction of the shape model starts with the following steps.

a) Translation: In this process, image is translated so that the mean of the objects points lies on the centroid of the object.
Landmarks_data_points= ({lmn.x1, lmny1}, {lmn.x2, lmny2}...),
The mean of these points is Xmean = ($lmn.x_1$+ $lmn.x_2$ + $lmn.x_{i)}$/I,Ymean = ($lmn.y_1$+ $lmn.y_2$ + $lmn.y_{i)}$/i
Translate, such that the mean is translated into origin (lmn.x.lmn.y) -> ($lmn.x_{1-}$ $Xmean.x_{1,}$ $lmn.y_{1-}$ $Ymean.y_{1)}$
b) Scaling:Likewise, the scale isremoved by re-scaling the object so that root mean square distance from the point to be translated to origin 1.
S = mean (lmn.x- $Xmean.x_1$)/ (lmn.y- $Ymean.y_1$),this division will make the scale 1 when the point coordinates are divided by the Bone part image object scale.
c) Rotation: Initially it involves computing the angle to center of all points and then subtracting the mean angle to build new data points that have same rotation.

3.3. Formulating the shape of the bone part: In this function the texture of the image is wrapped and aligned using Piecewise linear transformation.

3.4. Representing and measuring the variation of each segment of bone part:

Step 1: Compute the mean and variance of the data.

Step 2: Compute eigenvectors; retain 98% of the variation.

Step 3: Define matrix U from the top t eigenvectors.

Step 4: Approximation.

3.5. Learning Shape model and allowing variation of each Bone part segment.

4. Evaluation of the segmentation algorithm

For the complete validation of the segmentation, recall and precision metrics were computed with twenty one images as shown in Table 1 and 2.

The Recall values Table 1&2 show the proportion of correct and valid images segmentation that can be used in bone age process made by the system out of full database of twenty one segmented images.

Table 1.Distal phalanges Segmentation Evaluation Figure

Random Set ID	Number of Images	Image Type	Precision	Recall
1	2	Distal	2/2 =1	2/21=0.095
2	3	Distal	2/3 =0.66	2/21=0.095
3	4	Distal	3/4 =0.75	3/21=0.144
4	5	Distal	5/5= 1	5/21=0.239
5	6	Distal	6/6=1	6/21=0.288
6	5	Distal	4/5=0.8	6/21=0.298
		Average	**0.8683**	**0.1867**

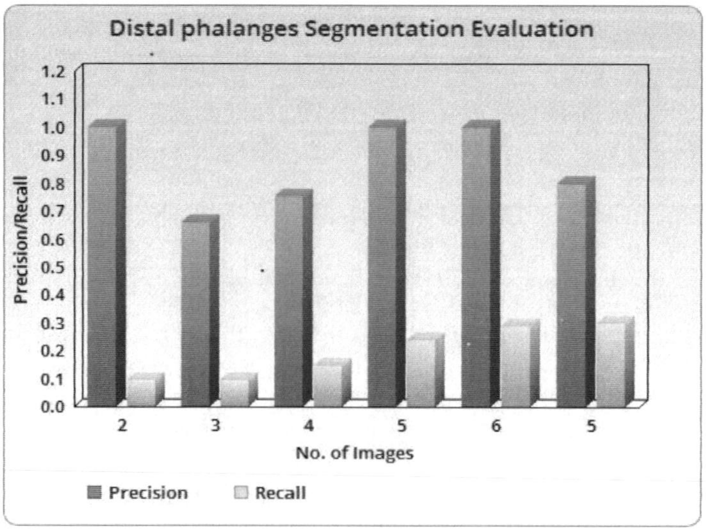

Figure 3. Graph for Distal phalanges Segmentation Evaluation

Table 2. Meta –Carpal Segmentation Evaluation

Random Set ID	Number of Images	Image Type	Precision	Recall
1	2	Meta Carpel	2/2 = 1	2/21=0.095
2	3	Meta Carpel	3/3=1	3/21=0.144
3	4	Meta Carpel	3/4= 0.75	3/21=0.144
4	5	Meta Carpel	3/5 =0.6	3/21=0.144
5	6	Meta Carpel	3/6 =0.5	3/21=0.144
6	5	Meta Carpel	5/5 = 1	5/21=0.238
		Average	**0.808**	**0.151**

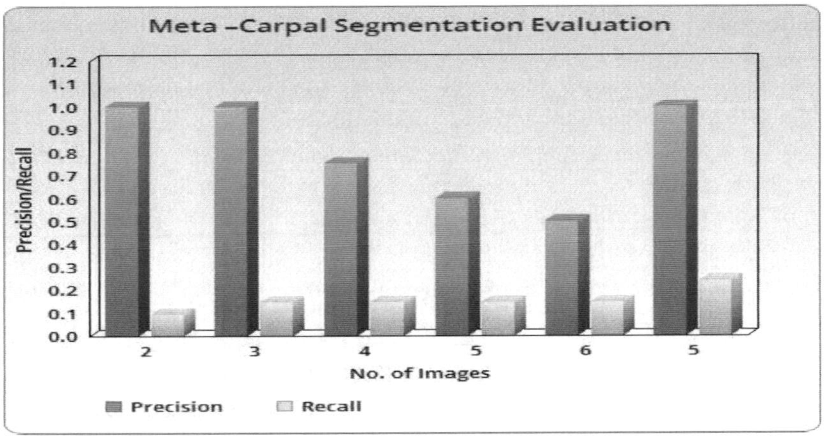

Figure4. Graph Representing Number of Images Selected from different sets

It is also found that high degree pre-selection (with help of medical expert) and pre-processing of the images have also lead to the minimization of complications involved identification of the hand bone development of children. The choice of retaining minimum number of Eigen values leads to construct fairly good model, hence leads to good segmentation output. Variation in hand structural positioning has lead to inaccuracies also. It can be seen from the table 1 & 2 the average precision remains above 0.86 in case of distal and 0.808 in case of metacarpal bones. The values of recall follow a similar pattern with respect to the precision values. The values of recall remain small when the evaluation is considered from full dataset. Care has been taken to build random sets of images called the evaluation set out of the segmented images. In total 42 images were segmented, 21 images for distal and 21 for Meta carpal bones.

5. Conclusion

This research work addresses methods and strategies that are employed for the segmentation of the bone parts from the X-ray image of full hand. These segmented images will be useful for the bone age assessment. Hence, the primary focus of the paper has been to build and efficient method of image ROI segmentation using custom models for each part. In real life conditions also, as mentioned in the introductory section that inter-rater differences make the bone age process vulnerable to many error that may have some legal and medical implication not expected .These days parents taking this bone age assessment simply to get prediction how much eventually their ward would be taller and see if that augured well for the competitive sports or just to find if there is some “constitutional growth and Pubertal delay (CGPD). The applications are many and not just restricted for detection bone diseases. In cases where precocious puberty occurs and advance bone formation take place then standard atlas, procedure and automated software may run into error. Therefore, some cases become typical in nature for the doctors and subject’s chronological age must also be considered for correct assessment. The algorithm works with few parameter tuning in each bone part model and the training is based on optimization methods with robust outcomes. The utilization of prior expected knowledge helps to build better models .The algorithms is relatively resistant to noise due to selection of high quality of images for training set. Then, the algorithm is adaptive in nature and optimally feasible for commercial applications of boneage assessment systems that may be built on cloud technologies. These models created here will reduce the manual work based on which researchers can build BAA in future.

6. References

[1] Vicente Gilsanz and Osman Ratib, "Digital Bone Age Atlas," in *Hand Bone Age.*: Springer, 2012, pp. 21-25.

[2] Marjan Mansourvar et al., "Automated bone age assessment: motivation, taxonomies, and challenges," *Computational and Mathematical Methods in Medicine*, vol. 2014, 2014, Histogram , Table 9.

[3] Emran Mohammad Abu Anas et al., "Automatic Segmentation of Wrist Bones in CT Using a Statistical Wrist Shape $+ $ Pose Model," *IEEE transactions on medical imaging*, vol. 35, no. 8, pp. 1789-1801, 2016.

[4] Darko Stern and Martin Urschler, "From individual hand bone age estimates to fully automated age estimation via learning-based information fusion," in *Biomedical Imaging (ISBI), 2016 IEEE 13th International Symposium on*, 2016, pp. 150-154.

[5] Jinwoo Seok, Jos, Michael DiPietro, and Anouck Girard, "Expert system for automated bone age determination," *Expert Systems with Applications*, vol. 50, pp. 75-88, 2016.

[6] Shreyas Simu and Shyam Lal, "A study about evolutionary and non-evolutionary segmentation techniques on hand radiographs for bone age assessment," *Biomedical Signal Processing and Control*, vol. 33, pp. 220-235, 2017.

[7] Rajitha Bakthula and Suneeta Agarwal, "Segmentation of bone pixels from EROI Image using clustering method for bone age assessment," in *SPIE Medical Imaging*, 2016, pp. 97890H--97890H, Bone Pixels.

[8] Pradipta Maji, "Advances in Rough Set Based Hybrid Approaches for Medical Image Analysis," in *Springer*, 2017, pp. 25--33.

[9] Shan and Li, Laquan and Choi, Wookjin and Kang, Min Kyu and D'Souza, WD and Lu, Wei Tan, "daptive region-growing with maximum curvature strategy for tumor segmentation in (18) F-FDG PET.," *Physics in medicine and biology*, vol. 62, no. 13, p. 5383, July 2017.

[10] Chi-Wen Hsieh, Tzu-Chiang Liu, Tai-Lang Jong, and Chui-Mei Tiu, "A fuzzy-based growth model with principle component analysis selection for carpal bone-age assessment," *Medical \& biological engineering \& computing*, vol. 48, no. 6, pp. 579-588, 2010, Fuzzy Method , Table 9.

[11] Guowei Gao, Chenglin Wen, Huibin Wang, and Lizhong Xu, "Fast Multiregion Image Segmentation Using Statistical Active Contours," *IEEE Signal Processing Letters*, vol. 24, no. 4, pp. 417-421, 2017.

[12] Qi Ge, Xiao-Yuan Jing, Fei Wu, Jingjie Yan, and Hai-Bo Li, "Unsupervised Joint Image Denoising and Active Contour Segmentation in Multidimensional Feature Space," *Mathematical Problems in Engineering*, vol. 2016, 2016, Countors.

[13] Yuanyao Lu and Tenghe Zhou, "Lip segmentation using localized active contour model with automatic initial contour," *Neural Computing and Applications*, pp. 1-8, 2017.

[14] Yusuke Shibayama et al., "Investigation of interfractional shape variations based on statistical point distribution model for prostate cancer radiation therapy," *Medical Physics*, vol. 44, no. 5, pp. 1837-1845, 2017, Point Distribution Model.

[15] Nannan Wang, Xinbo Gao, Dacheng Tao, Heng Yang, and Xuelong Li, "Facial feature point detection: A comprehensive survey," *Neurocomputing*, 2017, Point Distribtion.

[16] Xianghu Ji, Lili Wang, Pan Tao, and Zhongliang Fu, "Landmark Selecting on 2D Shapes for Constructing Point Distribution Model," in *Chinese Conference on Pattern Recognition*, 2016, pp. 318-331, Point Distibution.

The Precision Exploration of Image Super-Resolution Applied into Motion Measurement

Ye Wei, Wenhui Hou, Wei Feng, Yi Jin, Chang'an Zhu

School of Engineering Science, University of Science and Technology of China, Hefei, People's Republic of China

E-mail: jinyi08@ustc.edu.cn, weiye128@mail.ustc.edu.cn

Abstract. This paper proposes a new evaluation parameter of image super-resolution reconstruction quality, which is used to determine the influence of super-resolution reconstruction on motion measuring precision. Based on this new evaluation parameter, we respectively employ different algorithms of super-resolution reconstruction with different reconstruction scales, then calculate the motion measuring precision using the corresponding super-resolution image, analyze the impact of image super-resolution technology on the precision of image motion measuring. The experimental results show that only part of the super-resolution algorithms can improve the measuring precision slightly, and prove that under the background of image motion measurement, image super-resolution reconstruction method has the potential to apply to the actual measurement.

1. Introduction

It's undeniable that the computer vision motion measurement has been increasingly popular and applied in recent years based on its non-contact measurement [1]. However, the computer vision measurement precision is mainly depends on the resolution of the camera. The camera with high resolution will bring the higher calibration accuracy. Namely, under the background of the same pixel error, the high accuracy of measurement can certainly be obtained with the high resolution image. However it's upsetting that the price of the high resolution camera is not proportional to the resolution multiple of camera but far higher than that of low resolution camera. Besides, high resolution image often take up more storage space. Therefore, the image super resolution (SR) has very import application value and has been widely used in varieties of fields such as HDTV, video surveillance, remote sensing and medical image processing [2].

Existing SR algorithms can be classified into three categories: interpolation-based, reconstruction-based, and learning-based methods. Interpolation-based approaches, such as bilinear and bicubic interpolation, are the most commonly used methods in practice to upscale images. However, they tend to produce ringing and jagged artifacts as well as over-smoothing the critical structures such as edges [3], lead to the image blurring. In order to solve the above drawbacks, Li [4] proposed NEDI model which utilizes the similarity of the covariance matrix between the LR image and corresponding HR one to interpolate the pixels along the edges. Getreuer [5 introduces contour stencils for estimating the image contours and designing an edge directed color interpolation method.

By building an imaging model of the LR image, reconstruction based algorithms consider the image transformation in imaging, and look for the optimal estimation of the true HR image [6]. They assume a simplified continuous imaging process and usually formulate the imaging process as a linear system. Among them, the frequency domain method is a pioneering approach [7]. The reconstruction based approaches do not exploit prior knowledge for potential effect improvement. What's more, the high-frequency information often can't be reconstructed well with a large SR scaling factor. Hence, the reconstruction-based algorithms are limited in improving image resolution [8].

Finally, the learning-based approaches have become popular recently [9]. The learning procedure is used to supply prior knowledge to guide the SR reconstruction. This method can overcome many defects of the reconstruction-based approaches. And it mainly via structure analysis of image patches [10]. Also based on the image patches, Yang realized super-resolution and obtained a nice result via sparse representation which was named ScSR [11]. Similarly, the convolution neural network (CNN) was applied in this method by Dong [12]. But this method pays the price of large memory consumption. And if an external training database is used, noticeable artifacts often exist in the results when the specific image content does not match well with the training database. An alternative is to acquire the training set from the image itself based on local self similarities (LSS) [13], which can reduce memory consumption and achieve good results. LSS is based on research on image statistics that suggests image patches can be well-represented as a sparse linear combination of elements from an appropriately chosen over-complete dictionary. What's more, the integration of several algorithms can make the result better, like the method named as A+ [14].

This paper tries to test the influence of the high-resolution image reconstructed from the low resolution image on the accuracy of the image motion measurement, especially in the tiny motion measurement by using the image super-resolution reconstruction algorithms. So we proposed a new evaluation parameter of image super-resolution reconstruction quality, and based on it, we investigate the effects in different reconstruction scales and shifts with different algorithms: Bicubic, Yang [13], A+ [14], Kim [10], ScSR [11] and SRCNN [12] by using the image in the database. The remainder of this paper is organized as follows. Section 2 explains the meaning of the new evaluation parameter, and gives the main process of the test. Section 3 presents the results of these tests. Finally, the conclusion is drawn in Section 4.

2. The new evaluation factor

There are mainly two evaluation factors for the quality of image super-resolution reconstruction which are Peak signal-to-noise ratio [15] and Structural SIMilarity [16]. Peak signal-to-noise ratio, often abbreviated as PSNR, is an engineering term for the ratio between the maximum possible power of a signal and the power of corrupting noise that affects the fidelity of its representation. Because the dynamic range of many signals are very wide, PSNR is usually expressed in terms of the logarithmic decibel scale. The Structural SIMilarity (SSIM) index is a method for measuring the similarity between two images. The SSIM index can be viewed as a quality measure of one image compared with the other image whose quality is regarded as perfect.

In order to preferably evaluate the precision of image motion measurement with image super-resolution reconstruction algorithms, we propose the precision of SR in motion measurement, abbreviated as PRVM. To get the PRVM value of a algorithm utilized in a image, we need to know the accurate value of the motion. So the image displacement transformation is applied to simulate the motion of the image. The specific process can reference below, as shown in figure 1.

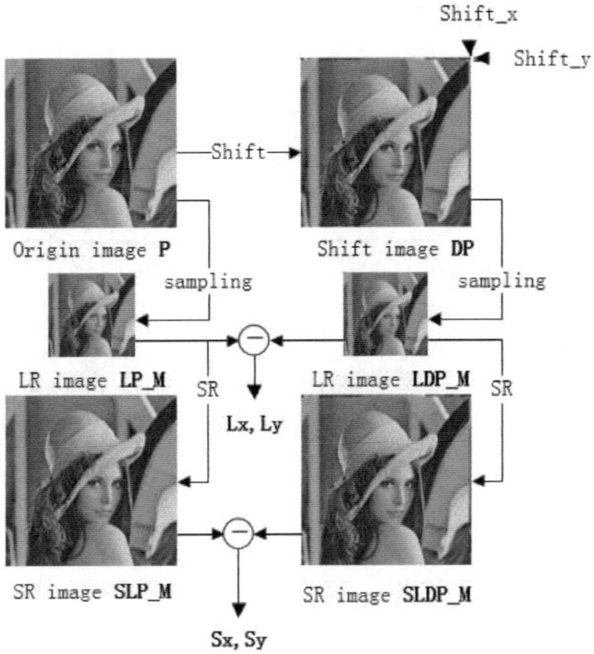

Figure 1. The solve process of PRVM

First of all, an original image P with high resolution is shifted by a presupposed value D, which is regarded as the image motion displacement in plane. And D = [shift_x shift_y]. In order to research the effect of SR algorithms in tiny motion, the value should be supposed smaller than the general one. The shift is implemented by applying a linear phase on its FT, thus we have assumed that the image is wrap around (features leaving one side of the window reappear on the opposite side) and that the image is band-limited (interpolated by a sinc function). Cross-correlation image registration by DFTs (both the matrix-multiply DFT and the zero-padded FFT) shared the same assumptions.

Then the original image P and the shifted image DP are both conducted the sampling process with scale M to reduce their resolutions. The two new images LP_M and LDP_M are calculated by the registration algorithm [17], which can obtain the shift values in the x and y directions respectively and can guarantee the accuracy of 0.001 pixel. So the shift values calculated from low resolution images are regarded as the measurement value in low resolution, written as Lx and Ly. After that, we reconstruct the images with the SR algorithms and got the high resolution images SLP_M and SLDP_M respectively. Similarly, we calculate the shift values Sx and Sy from the new high resolution images.

At last, the PRVM of the algorithm utilized in this original image with presupposed shift D can be presented as

$$PRVM = [|Sx - shift\_x| / shift\_x \qquad |Sy - shift\_y| / shift\_y] \tag{1}$$

And the measure error in low resolution images can be calculated by

$$Err = [|Lx \cdot M - shift\_x| / shift\_x \qquad |Ly \cdot M - shift\_y| / shift\_y] \tag{2}$$

Obviously, only when the PRVM is smaller than the Err, the algorithm can be considered effective.

3. The results of research

Take the picture ‘Lena.jpg’ for example, we conducted 7 SR reconstruction algorithms on it to test the effects. The reconstruction scales selected were 2, 3, 4 respectively. The original picture and the sampled pictures in different scales were shown in figure 2. The solution of ‘Lena.jpg’ is 512X512, in order to better represent the effect, we narrowed all the pictures in 3 times in the following paper.

(a) Original 'Lena.jpg' (b) scale=2 (c) scale=3 (d) scale=4

Figure 2. The original image and its sampled images in scale 2, 3, 4.

We set 3 shifts to simulate the different motions, shift 1 = [-5.7534 6.2157], shift 2 = [7.4702 -2.0951], shift 3 = [1.1452 8.3576]. The values in x and y directions of Shift 1 are close while those of shift 2 and shift 3 were largely different in two directions. The shifted images are shown in figure 3.

(a) shift 1 (b) shift2 (c) shift3

Figure 3. The shifted images in different shift values

The SR algorithms selected are Bicubic, Yang [13], A+ [14], Kim [10], ScSR [11] and SRCNN [12]. And the reconstructed images of the sampled image with scale 2 are presented below in figure 4.

Bicubic Yang [13] A+ [14] Kim [10] ScSR [11] SRCNN [12]

Figure 4. the reconstructed images by the figure 2(b) in different SR algorithms

3.1. The results of different scales and shifts

The first content we test is the effect of SR reconstruction scale on the measure precision. The reconstructed images of shift 1 in different scales are shown in figure 5.

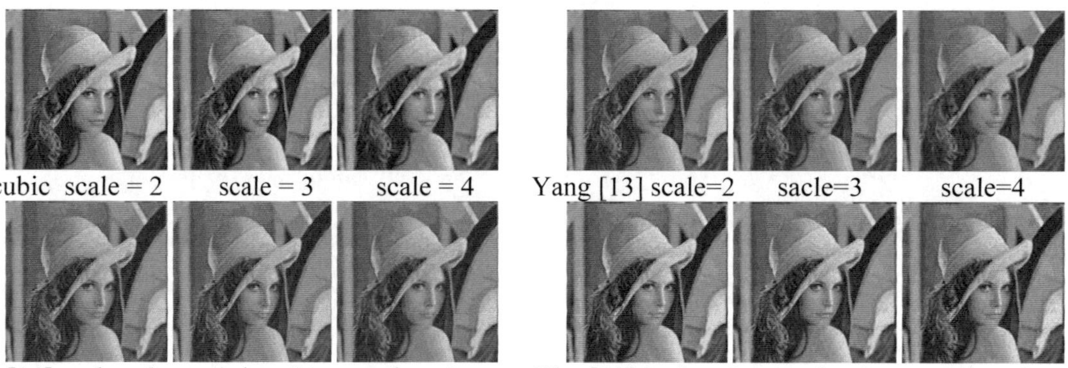

Bicubic scale = 2 scale = 3 scale = 4 Yang [13] scale=2 sacle=3 scale=4

A+ [14] scale = 2 scale = 3 scale = 4 Kim [10] scale = 2 scale =3 scale = 4

IOP Conf. Series: Journal of Physics: Conf. Series **933** (2017) 012005 doi:10.1088/1742-6596/933/1/012005

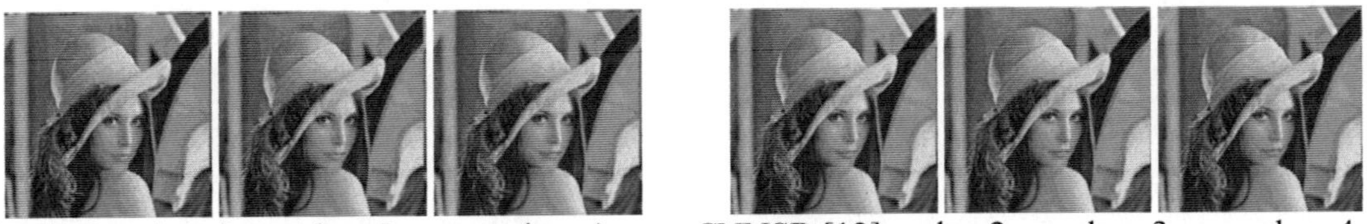

Figure 5. The SR images of shift 1 in different scales and algorithms

The values of Errs with shift 1 in three scales are [0.06% 0.23%], [0.41% 0.98%] and [0.25% 0.01%] respectively. And the PRVM values of different algorithms in different scales are calculated in table 1, in which the effective algorithm results are bold.

Table 1. The PRVM values of different algorithms and scales with shift 1

	Bicubic	Yang	A+	Kim	ScSR	CNNSR
2X	0.13%	2.06%	1.41%	0.09%	0.28%	1.16%
	0.27%	2.30%	1.85%	0.42%	0.42%	1.46%
3X	0.64%	2.79%	1.67%	0.56%	0.92%	1.37%
	0.85%	3.64%	2.62%	**0.78%**	**0.48%**	2.07%
4X	0.34%	5.08%	3.58%	**0.01%**	0.27%	2.31%
	0.34%	4.94%	3.27%	0.34%	1.39%	2.19%

Similarly, the PRVM values with shift 2 and shift 3 are presented in table 2 and table 3 respectively.

Table 2. The PRVM values of different algorithms and scales with shift 2

	Bicubic	Yang	A+	Kim	ScSR	CNNSR
2X	0.49%	1.23%	0.71%	0.29%	0.51%	-0.69%
	0.25%	5.89%	4.74%	1.48%	0.64%	4.11%
3X	1.33%	2.02%	1.09%	**0.98%**	1.58%	**1.03%**
	4.48%	7.43%	4.44%	3.89%	8.11%	3.25%
4X	1.14%	2.47%	1.40%	0.29%	1.90%	1.12%
	2.98%	11.78%	9.21%	**0.37%**	3.86%	6.90%

Table 3. The PRVM values of different algorithms and scales with shift 3

	Bicubic	Yang	A+	Kim	ScSR	CNNSR
2X	1.58%	4.94%	2.84%	**0.42%**	2.43%	3.71%
	0.51%	1.61%	1.47%	0.39%	0.58%	1.17%
3X	**3.06%**	13.72%	13.56%	7.89%	9.40%	8.01%
	1.59%	1.68%	**0.44%**	1.57%	2.41%	**0.63%**
4X	13.16%	16.24%	13.03%	13.33%	16.41%	9.67%
	0.49%	3.34%	2.67%	0.52%	0.93%	1.95%

The data shows that the image measurement has a certain accuracy under the low resolution in image measurement, but it is arbitrary to conclude that the SR reconstruction algorithms can improve the precision of the measurement only from the above data. In order to better compare the impact on the results of different reconstruction scales, we classify the above data according to the reconstruction scale, and accumulate the effective times under the same reconstruction scale. The results are shown in table 4.

Table 4. The effective times of different algorithms in different scales for 'Lena.jpg'

	Bicubic	Yang	A+	Kim	ScSR	CNNSR
2X	0/6	0/6	0/6	1/6	0/6	0/6
3X	2/6	0/6	1/6	2/6	1/6	2/6
4X	0/6	0/6	0/6	2/6	0/6	0/6

In table 4, each algorithm is conducted 6 times measuring calculations for each scale. And what can be seen from table 4 is that, for image 'Lena. JPG', all kinds of SR reconstruction algorithms can't improve the results obviously, including Kim [12] algorithm is slightly better than other algorithms. The reconstruction scale does not evidently effect on the reconstruction of the algorithm.
Similarly, the classification of the shift is implemented, and the corresponding results of each shift in both directions are shown in table 5.

Table 5. The effective times of different algorithms in different shifts for 'Lena.jpg'

	Bicubic	Yang	A+	Kim	ScSR	CNNSR
Shift 1	0/3	0/3	0/3	1/3	0/3	0/3
	1/3	0/3	0/3	1/3	1/3	0/3
Shift 2	0/3	0/3	0/3	1/3	0/3	1/3
	0/3	0/3	0/3	1/3	0/3	0/3
Shift 3	1/3	0/3	0/3	1/3	0/3	0/3
	0/3	0/3	1/3	0/3	0/3	1/3

From table 5, we can get that, just for image 'Lena. JPG', the motion of the different direction doesn't affect the SR reconstruction's influence on the accuracy of measurement. In general, it is not adequate to get the effective conclusion merely rely on the results of one image, so we need to test more image data.

3.2. The results of different images

In order to guarantee the validity of the results, we choose 4 images from BSD500 database to conduct this test which are: '8086.JPG', '64061.JPG', '257098.JPG' and '372019.JPG'. Via accumulating all the images results, the classification results according to the reconstruction scale are shown in table 6.

Table 6. The effective times of different algorithms in different scales

	Bicubic	Yang	A+	Kim	ScSR	CNNSR
2X	7/30	5/30	8/30	11/30	10/30	10/30
3X	10/30	6/30	10/30	11/30	10/30	12/30
4X	10/30	5/30	8/30	15/30	8/30	15/30

From table 6, we can get that with the increasing of the reconstruction scale, the image reconstructed by SR algorithms become more suitable than the original low resolution image for image measurement, especially the effects of algorithm Kim [10] and CNNSR [12] are more noticeable. According to this trend, it is presumable that the image of SR reconstruction can partly improve the precision of image measurement in the condition of high reconstruction scale.
Statistics of all images results of different motion direction are shown in table 7.

Table 7. The effective times of different algorithms in different shifts

	Bicubic	Yang	A+	Kim	ScSR	CNNSR
Shift 1	2/15	3/15	4/15	5/15	1/15	6/15
	5/15	0/15	1/15	6/15	6/15	7/15
Shift 2	6/15	7/15	11/15	6/15	6/15	11/15
	4/15	2/15	1/15	4/15	2/15	3/15
Shift 3	4/15	1/15	3/15	8/15	6/15	3/15
	6/15	2/15	6/15	7/15	6/15	7/15

It's very easy to find the difference between row 'Shift 2' and row 'Shift 3' in table 7. And it illustrates that the SR reconstruction images with a large shift is superior to the small one in the displacement measurement. Therefore, in the context of tiny motion measurement, the applicability of the image SR reconstruction is limited.

For different images, the influence of the image with 6 SR reconstruction algorithms is explored. The results are shown in table 8.

Table 8. The effective times of different algorithms for different images

	Bicubic	Yang	A+	Kim	ScSR	CNNSR
Lena	2/18	0/18	1/18	5/18	1/18	1/18
8086	6/18	5/18	7/18	7/18	5/18	11/18
64061	6/18	5/18	6/18	7/18	7/18	9/18
257098	5/18	1/18	4/18	5/18	4/18	5/18
372019	8/18	5/18	8/18	13/18	10/18	11/18

As can be seen from the table 8, the applicability of SR algorithm is also different for different images. We can see that all sorts of algorithms are unsatisfactory for 'Lena.JPG', however the good results can be obtained for '372091.JPG'. Compare these two images, we found that the boundaries in the image '372091.JPG' is more obvious than 'Lena.JPG'. So we can speculate that the SR reconstruction algorithm can significantly optimize the image measurement for the image with clear edges.

In general, the SR reconstruction algorithm, can improve the precision of image measurement on a certain extent, and the performance of algorithm Kim [10] and CNNSR [12] is more outstanding.

4. Conclusions

In this paper, we sample and reconstruct the simulated motion images through the usage of image super-resolution reconstruction algorithm, and establish a new evaluation mechanism for the precision of image motion measurement. And we try to test the value of image super-resolution reconstruction in the field of image measurement. The results show that the super-resolution image reconstruction can only improve the motion measurement precision of the image in the image measurement on a certain extent.

The results show that the reconstruction images do not affect the accuracy of measurement significantly within a certain range of image reconstruction magnification, but there is a tendence that the effects would be better when the reconstruction scale increased; The main motion direction don't effect on the accuracy of measurement. The error for small displacement measurement is larger than the big displacement, which illustrates that the SR reconstruction is not suitable for micro motion measurement; The influence of all kinds of super-resolution reconstruction algorithm on image measurement precision is random. And the edge information contrast in the image has great effect. The image with big edge information contrast is more suitable for SR reconstruction algorithm.

Among all the compared algorithms, the effects of algorithm Kim [10] and SRCNN [12] are better. In general, it requires further processing and optimization if you want to apply image super-resolution algorithm to image motion measurement process.

For the practical application of the SR reconstruction on the motion measurement, we plan to improve the preprocessing process and find the best parameters of the applied SR reconstruction algorithms. The next work will be better in real measurement.

5. References

[1] Guo, J., Zhu, C., Lu, S., Zhang, D. and Zhang, C. 2016. Vision-based measurement for rotational speed by improving Lucas – Kanade template tracking algorithm. Applied Optics 55, 25 (2016), 7186. DOI= http://dx.doi.org/10.1364/ao.55.007186.

Liu, B., Zhang, D., Guo, J. and Zhu, C. 2016. Vision-based displacement measurement sensor using modified Taylor approximation approach. Optical Engineering 55, 11 (2016), 114103. DOI= http://dx.doi.org/10.1117/1.oe.55.11.114103.

[2] A. Lapini, F. Argenti, A. Piva, L. Bencini. 2014. Comparison of super-resolution methods for quality enhancement of digital biomedical images. 2014 8th International Symposium on Medical Information and Communication Technology (ISMICT) (2014). DOI= http://dx.doi.org/10.1109/ismict.2014.6825243.

[3] P. Thevenaz, T. Blu, M. Unser. 2000. Image Interpolation and Resampling. Handbook of Medical Imaging (2000), 393 – 420. DOI= http://dx.doi.org/10.1016/b978-012077790-7/50030-8.

[4] X. Li, M.T. Orchard. 2000. New edge directed interpolation. Proceedings 2000 International Conference on Image Processing (Cat. No.00CH37101) (2000). DOI= http://dx.doi.org/10.1109/icip.2000.899369.

[5] P. Getreuer. 2011. Contour Stencils: Total Variation along Curves for Adaptive Image Interpolation. SIAM Journal on Imaging Sciences 4, 3 (2011), 954 – 979. DOI= http://dx.doi.org/10.1137/100802785.

[6] Z. Lu, C. Wu, D. Chen, et al.. 2014. Overview on image super resolution reconstruction. Control and Decision Conference (CCDC). IEEE. pp. 2009 – 2014. DOI= http://dx.doi.org/10.1109/CCDC.2014.6852498.

[7] R. Khan, R. Sablatnig, A. Bais, et al. 2011. Comparison of reconstruction and example-based Super-Resolution. 2011 7th International Conference on Emerging Technologies (2011). DOI= http://dx.doi.org/10.1109/icet.2011.6048463.

[8] S. Baker, T. Kanade. Limits on super-resolution and how to break them. EEE Trans. Pattern Anal. Mach. Intell. 24 (9) (2002) 1167 – 1183. DOI= http://dx.doi.org/10.1109/cvpr.2000.854852.

[9] Q. Ning, K. Chen, L. Yi, et al. 2013. Image Super-Resolution Via Analysis Sparse Prior. IEEE Signal Processing Letters 20, 4 (2013), 399 – 402. DOI= http://dx.doi.org/10.1109/lsp.2013.2242198.

Y. Zhang, J. Liu, W. Yang, Z. Guo. 2015. Image super-resolution based on structure-modulated sparse representation. IEEE Trans. Image Process. 24 (9) (2015) 2797 – 2810. DOI= http://dx.doi.org/10.1109/TIP.2015.2431435.

[10] C. Kim, K. Choi, J.B. Ra. 2013. Example-Based Super-Resolution via Structure Analysis of Patches. IEEE Signal Processing Letters 20, 4 (2013), 407 – 410. DOI= http://dx.doi.org/10.1109/lsp.2013.2249514.

[11] Jianchao Yang, John Wright, Thomas S. Huang, and Yi Ma. 2010. Image Super-Resolution Via Sparse Representation. IEEE Transactions on Image Processing 19, 11 (2010), 2861 – 2873. DOI= http://dx.doi.org/10.1109/tip.2010.2050625.

[12] Dong C., Loy C.C., He K., Tang X. 2014. Learning a Deep Convolutional Network for Image Super-Resolution. European Conference on Computer Vision (ECCV 2014). Lecture Notes

in Computer Science, vol 8692. Springer, Cham. DOI= https://doi.org/10.1007/978-3-319-10593-2_13.

[13] R.Zeyde, M.Elad, M.Protter. 2012. On Single Image Scale-Up Using Sparse-Representations. Curves and Surfaces Lecture Notes in Computer Science (2012), 711 - 730. DOI= http://dx.doi.org/10.1007/978-3-642-27413-8_47.

C.Yang, J.Huang, M.Yang. 2011. Exploiting Self-similarities for Single Frame Super-Resolution. Computer Vision - ACCV 2010 Lecture Notes in Computer Science (2011), 497 - 510. DOI= http://dx.doi.org/10.1007/978-3-642-19318-7_39.

[14] R. Timofte, V. De Smet, and L. Van Gool. 2015. A : Adjusted Anchored Neighborhood Regression for Fast Super-Resolution. Computer Vision -- ACCV 2014 Lecture Notes in Computer Science (2015), 111 - 126. DOI= http://dx.doi.org/10.1007/978-3-319-16817-3_8.

[15] Q. Huynh-Thu and M. Ghanbari. 2008. Scope of validity of PSNR in image/video quality assessment. Electronics Letters 44, 13 (2008), 800. DOI= http://dx.doi.org/10.1049/el:20080522.

[16] Z. Wang, A.c. Bovik, H.r. Sheikh, and E.p. Simoncelli. 2004. Image Quality Assessment: From Error Visibility to Structural Similarity. IEEE Transactions on Image Processing 13, 4 (2004), 600 - 612. DOI= http://dx.doi.org/10.1109/tip.2003.819861.

[17] Manuel Guizar-Sicairos, Samuel T. Thurman, and James R. Fienup. 2008. Efficient subpixel image registration algorithms. Optics Letters 33, 2 (October 2008), 156. DOI= http://dx.doi.org/10.1364/ol.33.000156.

Acknowledgments

This work was supported by The National Basic Research Program (973 Program) (Grant nos.2014CB049500), the National Science Foundation of China (Grant nos.51605464) and the Anhui Provincial Natural Science Foundation (Grant nos.1508085QE83).

IOP Conf. Series: Journal of Physics: Conf. Series **933** (2017) 012006 doi:10.1088/1742-6596/933/1/012006

Automatic Detection of Welding Defects using Deep Neural Network

Wenhui Hou, Ye Wei, Jie Guo, Yi Jin, Chang'an Zhu

School of Engineering Science, University of Science and Technology of China, Hefei, People's Republic of China

E-mail: jinyi08@ustc.edu.cn, hwh303@mail.ustc.edu.cn

Abstract. In this paper, we propose an automatic detection schema including three stages for weld defects in x-ray images. Firstly, the preprocessing procedure for the image is implemented to locate the weld region; Then a classification model which is trained and tested by the patches cropped from x-ray images is constructed based on deep neural network. And this model can learn the intrinsic feature of images without extra calculation; Finally, the sliding-window approach is utilized to detect the whole images based on the trained model. In order to evaluate the performance of the model, we carry out several experiments. The results demonstrate that the classification model we proposed is effective in the detection of welded joints quality.

1. Introduction

With the increasing requirement for the quality of equipment, radiographic testing as one of the oldest non-destructive testing (NDT) is commonly used in the detection of welded joints quality in many industrial fields such as the nuclear, chemical and aeronautical [1]. The real-time detection and automatic identification for welding defects from the digitized gray images become the focus of NDT research. In the last three decades, several efforts have been devoted to the development of such a system which can detect and classify the welding defects automatically using X-ray images.

D Mery [2] proposed a scheme for detecting weld defects from digitized film images based on three steps: segmentation, texture features extraction and classification. The supreme precision of the system achieved 90.91%. Silva et al. [3] described a non-linear classifier based on artificial neural networks (ANNs) compared with the linear classifiers. A criterion of neural relevance was applied to evaluate the capacity of the classifier studied by the geometric features in order to prove that the quality of the features extracted is more important than the quantity of the features. And Jayendra Kumar et al. [4] extracted the texture features in different direction for several spatial pixel distances for classifying the different flaws and obtained an overall classification accuracy of 86.1% by using artificial neural network (ANN). In another work [5], they integrated texture and geometrical feature, and obtained an accuracy of 87.34%. While Gang Wang and Liao [6] implemented an automatic computer-aided identification system. Twelve numeric features extracted were sent into two classifiers: fuzzy k-nearest neighbor (fuzzy K-NN) and multi-layer perceptron (MLP) neural networks for classifying different flaw types. Mohamed S. Juan Zapata [7] described an adaptive-network-based fuzzy inference system (ANFIS) for recognizing welding defects in radiographic images. In this work, 12 geometrical features is chosen as the input of ANFIS. Ioannis Valavanis [8] used sequential backward selection (SBS) method to select useful features for reducing the computational burden, and they compared the accuracies of three different classifiers.

These systems mainly rely on three steps : digital image processing, feature extraction, and pattern recognition. The image processing is designed to locate the weld region and segment the potential defects from the x-ray image. Miguel Carrasco and Domingo Mery [9] directly used digital image processing techniques including noise attenuation filters, edge detection techniques, the watershed transform, and the distance transform to segment welding defects without the later steps. Nevertheless, the quantitative evaluation of the system performance is more complex. In the phase of feature extracting, the texture and geometrical features were commonly used. Usually these features are numerous and the calculations are complex. In addition, not all the features are sensitive to classification, thus the feature selection relying on artificial selection, sequential forward selection (SFS), multi-scale wavelet packet (MWP), Hough transform or principal components analysis (PCA) is essential in order to reduce the dimension of the features.

Nowadays, modern computer vision techniques, such as deep learning [10] and sparse coding [11] are providing a new way in detecting the object automatically from optical images. In this paper, an automatic classification model based on deep neural network is proposed. It implements unsupervised learning for intrinsic feature representation by using several sparse auto-encoders (SAEs) instead of calculating the texture or geometrical features and extra feature selection. Then the softmax regression with supervised learning is used to fine-tune the deep network. And the proposed model is trained and tested by the x-ray image patches. In addition, we utilize the trained model to inspect the whole image and find the defects region. And the results demonstrate that the classification model we proposed is enough effective in the detection of welded joints quality.

2. Deep neural network

2.1. sparse auto-encoder

Sparse auto-encoder (SAE) network including input layer, hidden layer and output layer is mainly used for unsupervised learning [12]. The structure is shown in Figure 1.

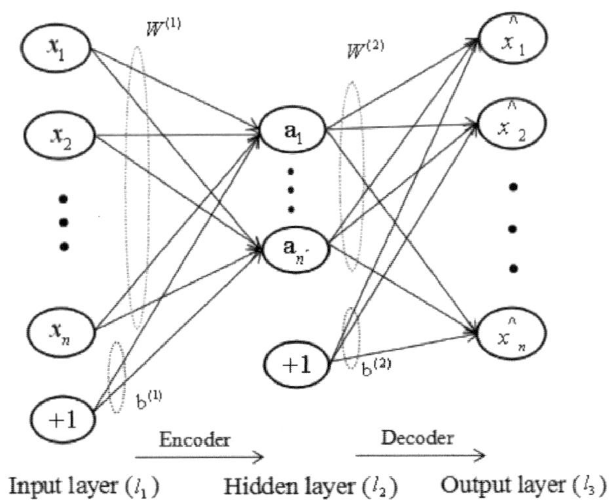

Figure 1. The structure of SAE

As is shown in figure 1, this network includes two parts. The one is encoder which is from input layer to hidden layer, $\mathbf{W}^{(1)}$ and $\mathbf{b}^{(1)}$ are the weight matrix and the bias of this part. The other one called decoder is from hidden layer to output layer, $\mathbf{W}^{(2)}$ and $\mathbf{b}^{(2)}$ are the corresponding parameters of this part.

The network tries to reconstruct the input vector in output layer, as is shown in Eq.1

$$\hat{\mathbf{x}} = h_{W,b}(\mathbf{x}) \approx \mathbf{x} \tag{1}$$

Where $\mathbf{x}$ is the input vector $\mathbf{x} \in R^{n \times 1}$. The cost function of SAE is given by

$$J(\mathbf{W},\mathbf{b}) = \{\frac{1}{m}\sum_{i=1}^{m}[\frac{1}{2}\left\|h_{W,b}(\mathbf{x}^{(i)}) - \mathbf{x}^{(i)}\right\|^2]\} + \frac{\lambda_1}{2}\sum_{l=1}^{2}\sum_{i=1}^{s_l}\sum_{j=1}^{s_{l+1}}(W_{ji}^{(l)})^2 + \beta\sum_{j=1}^{s_2}KL(\rho\|\hat{\rho}_j) \tag{2}$$

The first part of the function is the mean square term minimizing the error between the input data and the output data. $\mathbf{x}^{(i)}$ represent the ith sample, m is the amount of the samples. For each sample, $h_{W,b}(\mathbf{x}) = \sigma(\mathbf{W}^{(2)}\mathbf{a} + \mathbf{b}^{(2)})$ is the function of decoder, where $\mathbf{W}^{(2)} \in R^{n \times n'}$, $\mathbf{b}^{(2)} \in R^{n' \times 1}$. And $\mathbf{a}$ is the activation of hidden layer given by: $\mathbf{a} = \sigma(\mathbf{W}^{(1)}\mathbf{x} + \mathbf{b}^{(1)})$, i.e. the function of encoder. Where $\mathbf{W}^{(1)} \in R^{n' \times n}$, $\mathbf{b}^{(1)} \in R^{n' \times 1}$, σ is the sigmoid function, expressed as $(1+\exp(-x))^{-1}$, i.e. the activation function of encoder and decoder.

The second part of the function is the regularization term, also called the weight decay term avoiding over-fitting. Where λ_1 is the regularization factor, s_l represents the number of units in layer l which is equal to 3 in our network, and obviously, $s_1 = n$, $s_2 = n'$, $s_3 = n$. $W_{ji}^{(l)}$ is the weight between the i th unit in layer l and the j th unit in layer l+1.

The last part of the function is the sparse penalty term which impose a sparsity constraint on the hidden units, where β is the weight of this term. Kullback - Leibler divergence is given as Eq.3

$$KL(\rho\|\hat{\rho}_j) = \rho\log\frac{\rho}{\hat{\rho}_j} + (1-\rho)\log\frac{1-\rho}{1-\hat{\rho}_j} \tag{3}$$

Where ρ is the sparsity parameter which is usually small, $a_j^{(i)}$ represents the activation of the j th hidden unit for the i th sample. $\hat{\rho}_j$ is the average activation of all the samples given by $\hat{\rho}_j = \frac{1}{m}\sum_{i=1}^{m}[a_j^{(i)}]$. The L-BFGS and back propagation are used to train the network for minimizing the cost function.

2.2. *Softmax classifier*

Softmax classifier uses a supervised learning algorithm for multi-class classification problem. It works according to the probability belong to each category. The cost function is expressed as:

$$J(\boldsymbol{\theta}) = -\frac{1}{m}[\sum_{i=1}^{m} y^{(i)}\log h_{\boldsymbol{\theta}}(\mathbf{x}^{(i)}) + (1-y^{(i)})\log(1-h_{\boldsymbol{\theta}}(\mathbf{x}^{(i)}))] \tag{4}$$

Where $\boldsymbol{\theta} = [\boldsymbol{\theta}_1^T, \boldsymbol{\theta}_2^T, \ldots \boldsymbol{\theta}_k^T]^T \in R^{k \times (d+1)}$ is the parameters of this modal which is trained by minimizing the function. $\boldsymbol{\theta}_1, \boldsymbol{\theta}_2, \ldots \boldsymbol{\theta}_k \in R^{d+1}$, $\boldsymbol{\theta}_j^T$ is the jth row of matrix $\boldsymbol{\theta}$. $\mathbf{x}^{(i)} \in R^{d+1}$, k is the number of categories, $h_{\boldsymbol{\theta}}(x)$ represent the probability function belong to each category given by

$$h_{\boldsymbol{\theta}}(x^{(i)}) = \begin{bmatrix} p(y^{(i)}=1|\mathbf{x}^{(i)};\boldsymbol{\theta}) \\ p(y^{(i)}=2|\mathbf{x}^{(i)};\boldsymbol{\theta}) \\ \cdots \\ p(y^{(i)}=k|\mathbf{x}^{(i)};\boldsymbol{\theta}) \end{bmatrix} = \frac{1}{\sum_{j=1}^{k} e^{\boldsymbol{\theta}_j^T x^{(i)}}} \begin{bmatrix} e^{\boldsymbol{\theta}_1^T \mathbf{x}^{(i)}} \\ e^{\boldsymbol{\theta}_2^T \mathbf{x}^{(i)}} \\ \cdots \\ e^{\boldsymbol{\theta}_k^T \mathbf{x}^{(i)}} \end{bmatrix} \tag{5}$$

Where $y^{(i)} \in \{1,2,..k\}$, is the label of the input $\mathbf{x}$. $1/\sum_{j=1}^{k} e^{\theta_j^T x^{(i)}}$ is set to normalize the distribution. The cost function becomes Eq. 6 when a weight decay term is added.

$$J(\boldsymbol{\theta}) = -\frac{1}{m}\left[\sum_{i=1}^{m}\sum_{j=1}^{k} 1\{y^{(i)}=j\}\log\frac{e^{\theta_j^T \mathbf{x}^{(i)}}}{\sum_{l=1}^{k} e^{\theta_l^T \mathbf{x}^{(i)}}}\right] + \frac{\lambda_2}{2}\sum_{i=1}^{k}\sum_{j=1}^{d+1}\theta_{ij}^2 \tag{6}$$

where $1\{\}$ is the indicator function. λ_2 is the weight decay parameter, θ_{ij} present the weight between the j th unit of input layer and the i th unit of output layer. For each input vector, its category is judged by this classifier as the one whose probability is the maximum value of all categories.

2.3. Deep neural network model

In this part, a classification model based on deep neural network called SSAE was constructed by stacking several SAEs and a softmax classifier.

Firstly, we train SAEs one by one using the unlabeled samples for unsupervised learning. The hidden layer activities of the former SAE are used as the input to train the next SAE. And once the training of each SAE is finished, the decode layers of all SAEs are removed. The outputs of the last SAE are no other than the learned features. Then the Softmax classifier is connected to the rear of SAEs in order to finish the classification task. Lastly, the fine-tuning of the whole network using the labeled samples is a key step for better classification accuracy.

3. Experiment

3.1. The x-ray image set

The x-ray images used in this paper are obtained from a public database called GDXray [13]. The group 'welds'of the GDXray database is used in this paper. The group contains 88 images arranged in 3 series. The collection of these x-ray images is taken by BAM Federal Institute for Materials Research and Testing, Berlin, Germany. The samples of the proposed model for later learning are mainly selected from series W0001.

3.2. Pre-processing of the images

In this section, the weld region must be isolated from the rest of the elements in the x-ray images. In this way, the burden of identification is reduced. The process is developed in two steps. First, the gray scale image should be converted into a binary image according to an optimal threshold in order to separate an object's pixels from the background pixels. For this, the Otsu's method [14] is used which select the optimal threshold by minimizing the intraclass variance of the thresholded black and white pixels. Next, we crop an appropriate rectangular region from the original image according to the binary image for later sample acquisition.

An example of processed results is shown in figure 2.

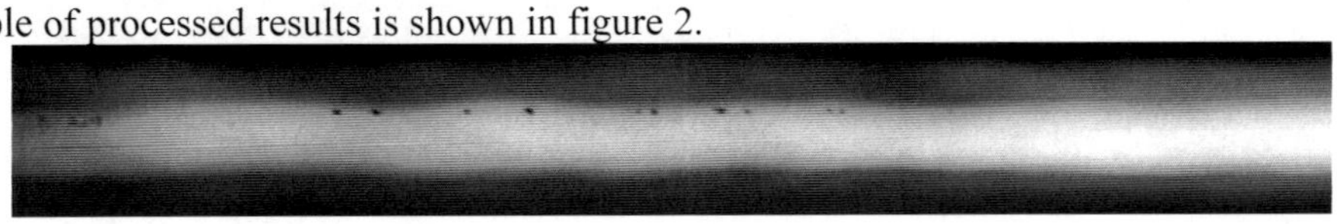

(a) original image

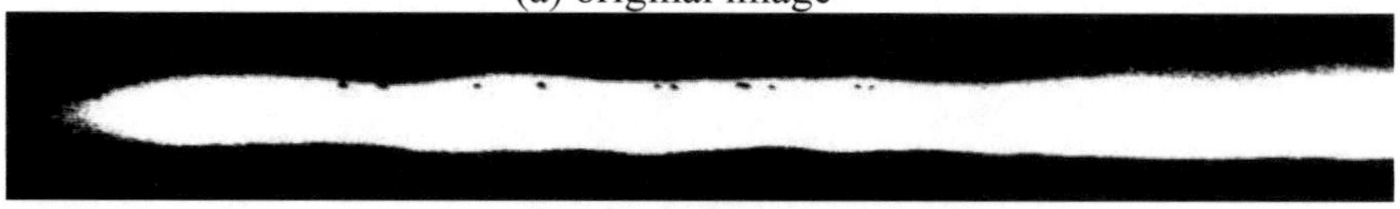

(b) binary image

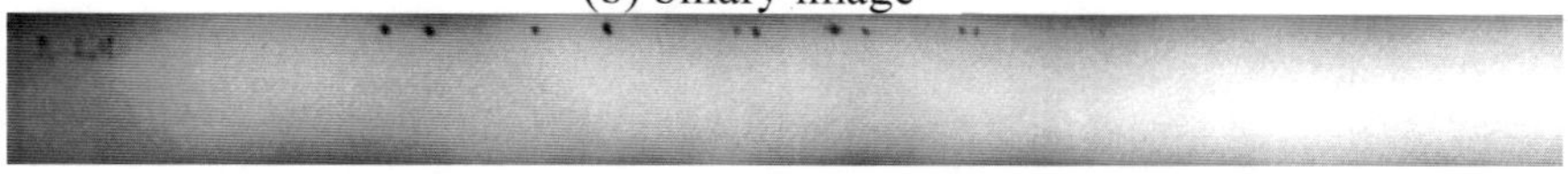

(c) weld region

Figure 2. The result of the processing

3.3. Classification and the detection via the proposed model

We crop patches with the size 32×32 pixels from the processed images as the deep learning sample set. These patches are labeled as 0 (no-defect) or 1 (defect) according to the annotations of bounding boxes. Several examples are shown in figure 3.

Several experiments are implemented including extracting the feature using SAE and examining the classification accuracy under different parameters of the model. In addition, we inspect the whole x-ray image using the slide-window method and recognized the defects through the proposed classification model. Figure 4 shows the overall flowchart of the experimental model.

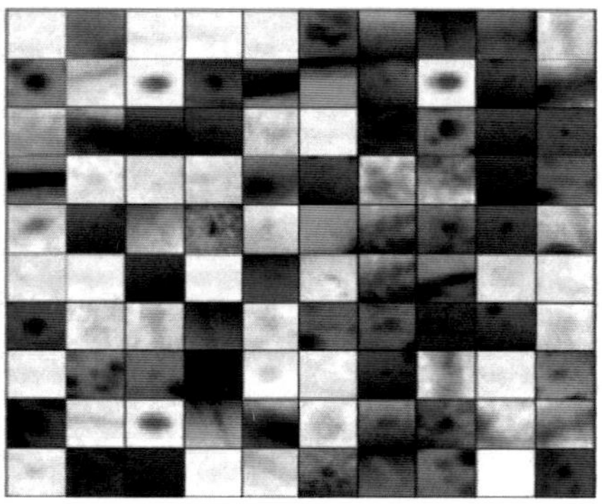

(a) patches with defects

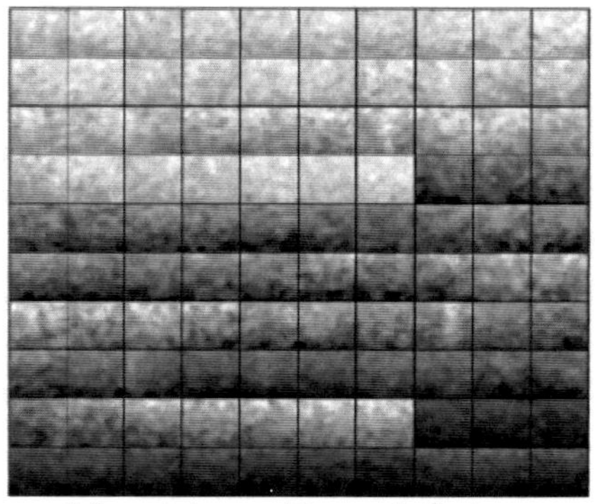

(b) patches with no-defects

Figure 3. The examples of patches

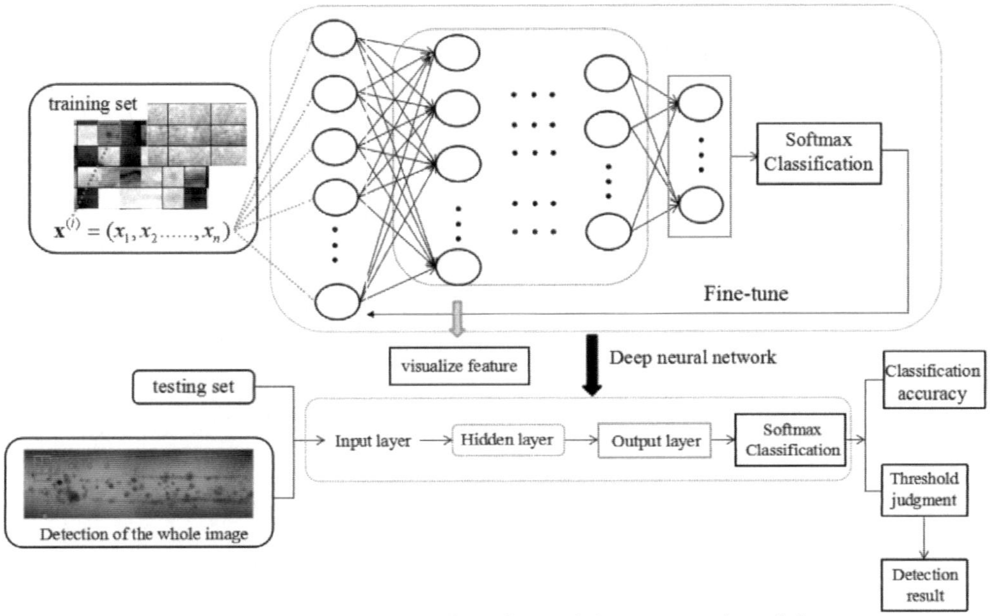

Figure 4. The overall flowchart of the proposed model

4. Experimental results

The performance of the proposed model is discussed through above-mentioned experiments. Next, the results of these experiments are shown in this section.

4.1. The result of unsupervised learning

In order to visualize the bases of image from single SAE, we choose 400 patches randomly from the sample set as the input of SAE. The features learned presenting different orientations and locations are shown in figure 5.

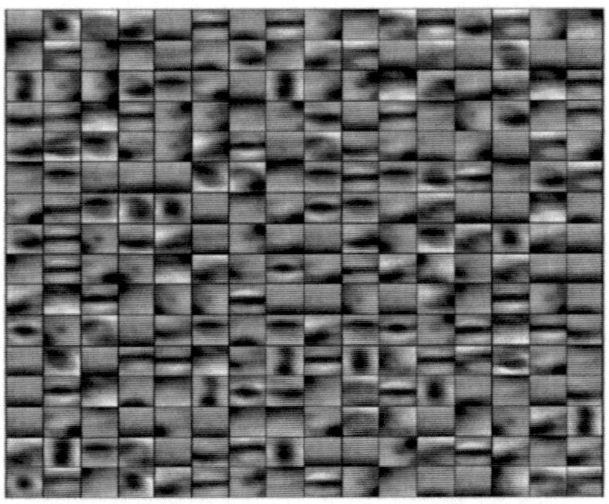

Figure 5. The features learned from SAE

4.2. Deep neural network parameter decision

In this section, in order to find the better parameters of deep neural network for classifying, we carry out several experiments to determine the number of hidden layers, the amount of neurons and the value of the sparsity parameter.

The influence of the number of network layers on classification performance is first examined. Two models with two hidden layers of 576 and 196 neurons and three hidden layers of 576, 196, 100 neurons are implemented, named as DL_net1 and DL_net2 respectively. In order to evaluate the classification performance of these two networks, we use 5-fold cross validation. So the training and testing are implemented 5 times. Table1 show the results of these 5 times testing for two models.

Table1. The results of testing

model	fold	P_r (%)	R_e (%)	$F1$ (%)	η (%)
DL_net1	1	84.82	90.00	87.33	87.63
	2	89.77	87.78	88.76	89.47
	3	90.34	88.33	89.33	90.00
	4	91.67	85.56	88.51	89.47
	5	83.77	88.89	86.25	86.58
	average	88.07	88.11	88.03	88.63
DL_net2	1	90.27	92.78	91.51	91.84
	2	87.50	89.44	88.46	88.95
	3	94.61	87.78	91.07	91.84
	4	89.77	87.78	88.76	89.47
	5	82.91	91.67	87.07	87.11
	average	89.01	89.89	89.37	89.84

In the table, the precision is presented as $P_r = \frac{TP}{TP+FP}$, the recall is defined as $R_e = \frac{TP}{TP+FN}$, the accuracy is defined as $\eta = \frac{TP+TN}{P+N}$, the F1 score is defined as $F1 = \frac{2 \times P_r \times R_e}{P_r + R_e}$. where TP is the number of defect patches correctly classified, TN is the number of no-defect patches correctly classified, FP is the number of no-defect patches classified as defect patches, FN is the number of defect patches classified as no-defect patches. We observe that accuracy of DL_net2 is about one percentage point higher than that of DL_net1.

The decision of the number of hidden neurons is another problem for the classification model. For this, we investigate the classification performance of two different networks which are DL_net21 with 576 - 196 -100 neurons and DL_net22 with 576 -196-50 neurons. The accuracies of two networks using hold-out testing versus the sparsity parameter changing from 0.04 to 0.12 are plotted in figure 6. The accuracy of DL_net21 achieve the maximum value of 90.9% at $\rho = 0.1$, while the one of DL_net22 achieve the maximum value of 90% at $\rho = 0.08$. Therefore, DL_net21 is adopted in the network adopted for later research and the value of the sparsity parameter selected is 0.1.

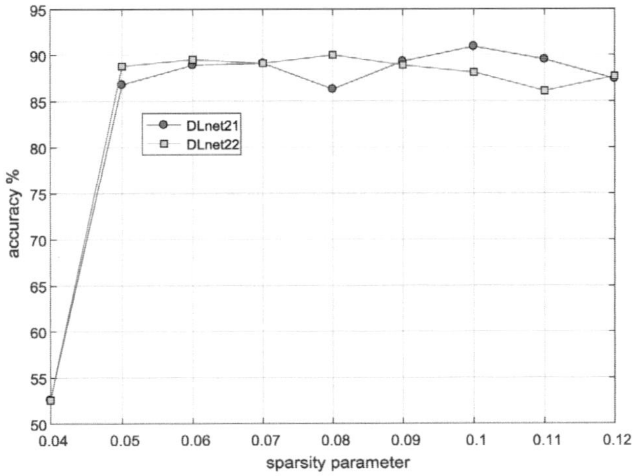

Figure 6. The accuracy versus sparsity parameter

4.3. Identification of defects for whole image

In this section, in order to validate the performance of our classification model, we implement the experiment on the whole x-ray image. For this, the sliding-window approach is adopted. The detection window with the size 32×32 pixels moves on the input image in both horizontal and vertical directions with step of 2 pixels. Every patch under the window is marked as "0" or "1" through the deep neural network. The pixels in the overlap of the detection windows are marked many times, and the mark values are accumulated. For each pixel, the mark value is varied between 1 and 256. When the value is more than an appointed threshold, the pixel is considered as defect. This process is called threshold judgement. The probability map for each pixel judged as a defect is shown in figure 7(a). The original image with the defect region marked is shown in figure 7(b). The results of other images are shown in figure 8.

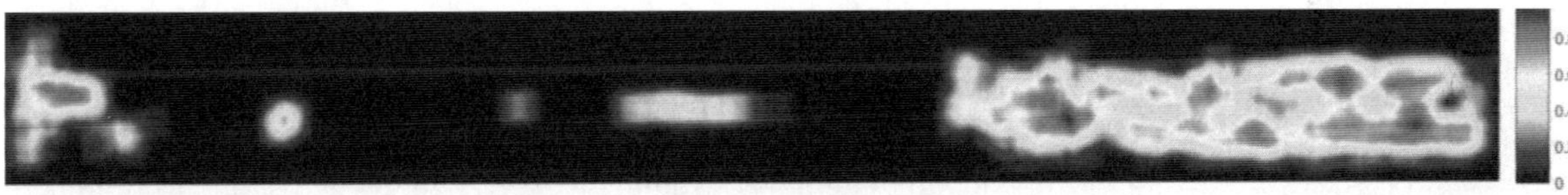

(a)

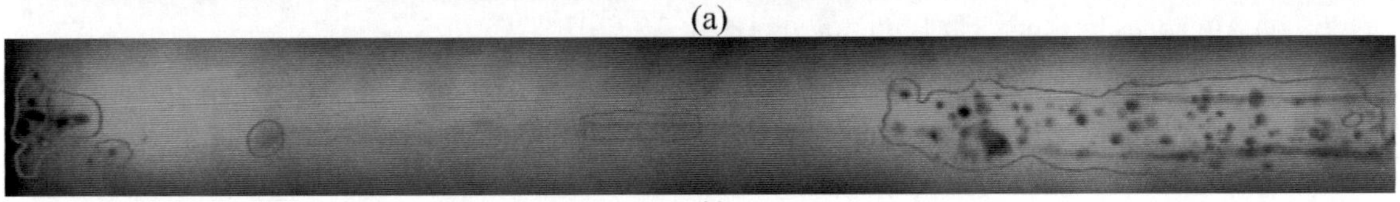

(b)

Figure 7. (a) The probability map for defects (b) The detection result

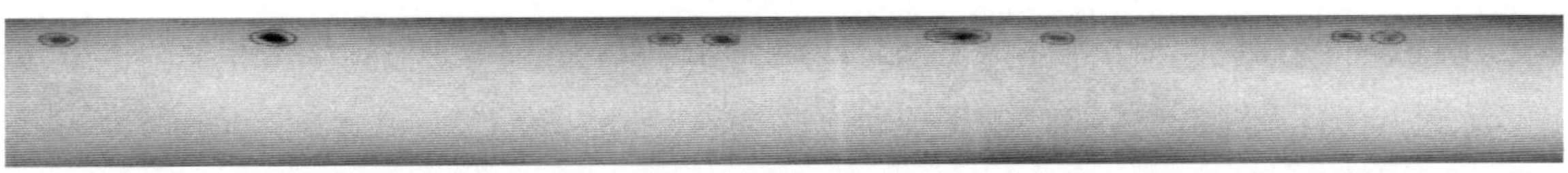

(a)

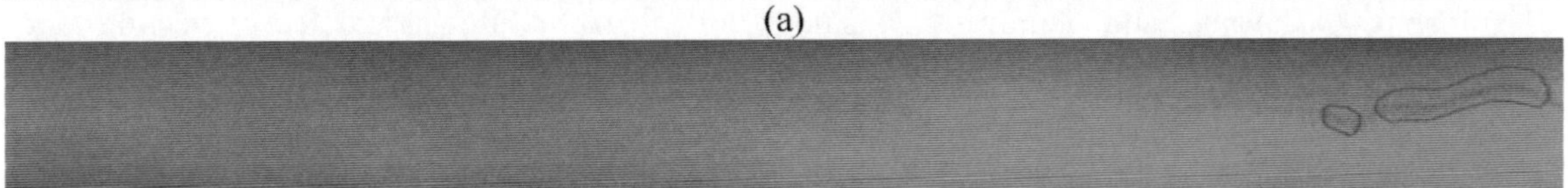

(b)

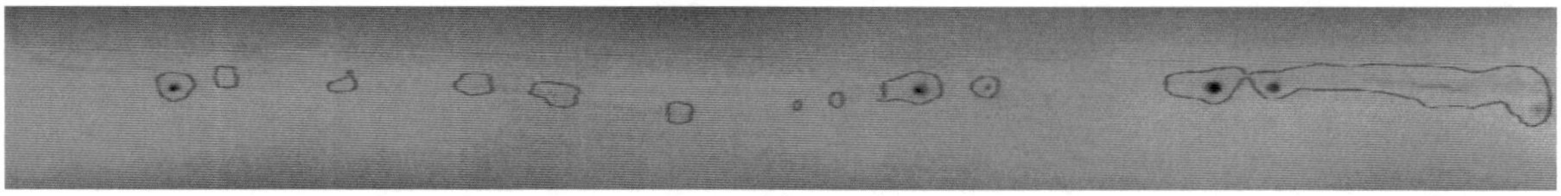

(c)

Figure 8. The detection results of other images

The results show that the proposed model can effectively recognize the defects of different types, such as gas hole, porosity, lack of penetration and crack, etc.

5. Conclusions

In this paper, we present an automatic detection model for weld defect based on deep neural network that extracts the intrinsic features of x-ray images. Several experiments have been conducted to validate our method and the influence of different parameters in the model on classification performance is also analyzed. Furthermore, we adopt the sliding-window approach to detect the whole x-ray image. The experiment results illustrate that the proposed model obtain a maximum classification accuracy rate of 91.84% (with 90.27% precision and 92.78% recall). For the whole images, the experiments of defect inspections are implemented based on the proposed model and the defects can be effectively recognized without distinguishing the specific types of the defects. In the future, we will try to classify the different weld defects of x-ray images.

6. References

[1] Vilar R, Zapata J, Ruiz R. C. 2016. An automatic system of classification of weld defects in radiographic images. NDT & E International 42, 5 (2009), 467 – 476.DOI= http://dx.doi.org/10.1016/j.ndteint.2009.02.004.

[2] D. Mery and M.A. Berti. 2003. Automatic detection of welding defects using texture features. Insight - Non-Destructive Testing and Condition Monitoring 45, 10 (2003), 676 – 681. DOI= http://dx.doi.org/10.1784/insi.45.10.676.52952.

[3] Romeu R.Da Silva, Luiz P. Calôba, Marcio H.s. Siqueira, and João M.a. Rebello. 2004. Pattern recognition of weld defects detected by radiographic test. NDT & E International 37, 6 (2004), 461 – 470. DOI: http://dx.doi.org/10.1016/j.ndteint.2003.12.004

[4] Jayendra Kumar, R.s. Anand, and S.p. Srivastava. 2014. Multi - Class welding flaws classification using texture feature for radiographic images. 2014 International Conference on Advances in Electrical Engineering (ICAEE) (2014). DOI= http://dx.doi.org/10.1109/icaee.2014.6838443

[5] Jayendra Kumar, R.S. Anand, and S.P. Srivastava. 2014. Flaws classification using ANN for radiographic weld images. 2014 International Conference on Signal Processing and Integrated Networks (SPIN) (2014). DOI= http://dx.doi.org/10.1109/spin.2014.6776938

[6] Gang Wang and T.warren Liao. 2002. Automatic identification of different types of welding defects in radiographic images. NDT & E International 35, 8 (2002), 519 – 528. DOI= http://dx.doi.org/10.1016/s0963-8695(02)00025-7

[7] Juan Zapata, Rafael Vilar, and Ramón Ruiz. 2010. An adaptive-network-based fuzzy inference system for classification of welding defects. NDT & E International 43, 3 (2010), 191 – 199. DOI= http://dx.doi.org/10.1016/j.ndteint.2009.11.002

[8] Ioannis Valavanis and Dimitrios Kosmopoulos. 2010. Multiclass defect detection and classification in weld radiographic images using geometric and texture features. Expert Systems with Applications 37, 12 (2010), 7606 – 7614. DOI= http://dx.doi.org/10.1016/j.eswa.2010.04.082

[9] Carrasco M A, Mery D. 2004. Segmentation of welding defects using a robust algorithm. Materials Evaluation,62(11) (2004), 1142-1147.

[10] Y. LeCun, Y. Bengio, and G. Hinton. 2015. Deep learning. Nature, 521(7553) (2015), 436 – 444, DOI= http://dx.doi.org/10.1038/nature14539

[11] Zehuan Yuan, Tong Lu, and Chew Lim Tan. 2017. Learning discriminated and correlated patches for multi-view object detection using sparse coding. Pattern Recognition 69 (2017), 26 – 38. DOI= http://dx.doi.org/10.1016/j.patcog.2017.03.033

[12] Ehsan Hosseini-Asl, Jacek M. Zurada, and Olfa Nasraoui. 2016. Deep Learning of Part-Based Representation of Data Using Sparse Autoencoders With Nonnegativity Constraints. IEEE Transactions on Neural Networks and Learning Systems 27, 12 (2016), 2486 – 2498. DOI= http://dx.doi.org/10.1109/tnnls.2015.2479223

Yan Zhang, Erhu Zhang, and Wanjun Chen. 2016. Deep neural network for halftone image classification based on sparse auto-encoder. Engineering Applications of Artificial Intelligence 50 (2016), 245 – 255. DOI= http://dx.doi.org/10.1016/j.engappai.2016.01.032

[13] D. Mery, V. Riffo, U. Zscherpel, G. Mondrag´on, I. Lillo, I. Zuccar, H. Lobel, and M. Carrasco. 2015. GDXray: The database of X-ray images for nondestructive testing. Journal of Nondestructive Evaluation, 34(4) (2015), 1 – 12. DOI = http://dx.doi.org/10.1007/s10921-015-0315-7.

[14] Nobuyuki Otsu. 1979. A Threshold Selection Method from Gray-Level Histograms. IEEE Transactions on Systems, Man, and Cybernetics 9, 1 (1979), 62 – 66. DOI= http://dx.doi.org/10.1109/tsmc.1979.4310076

Acknowledgments

This work was supported by The National Basic Research Program (973 Program) (Grant nos.2014CB049500), the National Science Foundation of China (Grant nos.51605464) and the Anhui Provincial Natural Science Foundation (Grant nos.1508085QE83).

IOP Conf. Series: Journal of Physics: Conf. Series **933** (2017) 012007 doi:10.1088/1742-6596/933/1/012007

Formalization of the Access Control on ARM-Android Platform with the B Method

Lu Ren, Wei Wang, Xiaodong Zhu, Yujia Man, Qing Yin

State Key Laboratory of Mathematic Engineering and Advanced ComputingZhengzhou, China, 450001

miranda.lu.rn@gmail.com

Abstract. ARM-Android is a widespread mobile platform with multi-layer access control mechanisms, security-critical in the system. Many access control vulnerabilities still exist due to the course-grained policy and numerous engineering defects, which have been widely studied. However, few researches focus on the mechanism formalization, including the Android permission framework, kernel process management and hardware isolation. This paper first develops a comprehensive formal access control model on the ARM-Android platform using the B method, from the Android middleware to hardware layer. All the model specifications are type checked and proved to be well-defined, with 75%of proof obligations demonstrated automatically. The results show that the proposed B model is feasible to specify and verify access control schemes in the ARM-Android system, and capable of implementing a practical control module.

1. Introduction

Recently, mobile devices have changed our lifestyle, like the BYOD (Bring Your Own Device), mobile payment, etc. The security of embedded system thus becomes increasingly important, where the access control policy plays a significant role. ARM-Android is a popular mobile system architecture equipped with multilevel access control mechanisms, such as the Android permission framework, kernel sandbox and hardware isolation. Although Android has some controversial features because of the application-level permission and incomplete access control policy, the current trusted execution technology, like ARM TrustZone, provides hardware isolation environment for the system. With the combination of hardware and softwareprotection prevalent in both industry and academia, most studies aim at the practical technique and system design. Xu et al. [1] implemented a TrustZone-based prototype system, KNOX [2] is a real-world product of Samsung, utilizing the TrustZone as the root of trust. However, there is no common access control model on the ARM-Android platform, supporting the practical technologies theoretically. As a formal method, B method has great advantages in system modeling and development, ensuring the high security of software systems.

This paper seeks to formalize the access control mechanism on ARM-Android platform, using the B language to describe the policies in different layers. A comprehensive formal model is constructed with the ability to visualize and auto-prove security properties progressively, which supports the model enhancement and consistency valuation as well. Based on Atelier B, the results illustrate that all the formal specifications are well-defined and 75% (535 out of 718) proof obligations are automatically proved, the rest to be proved interactively.

Our main contributions are:

i). We develop a formal Android permission model with the B method, specifying the inter-component communication and access control operation to maintain primary security properties.

ii). We design a process control model in the kernel and hardware layer, describing the process access management with the hardware-based isolation mechanism.

2. Background

2.1. Android overview

Android is an open source software stack including the Linux kernel, the middleware and the application layer[3].

Android Application. An Android application consists of various components, which should be instantiated by the system at runtime. Android isolates one application execution from others and provides the ICC (Inter Component Communication) mechanism. The AndroidManifest.xml file of an application gives some basic information, e.g., package attributes, an identifier, contained components and app-related permissions.

Android permission mechanism. Diverse permissions are applied for the resource protection, comprising the identifier, permission group and protection level. The permission-based security mechanism is enforced at both the application installation time and running time. When installed, an application requests a set of permissions to be granted by the user. At runtime, the system performs permission check on instantiated components and control ICCs. Android 6.0 (or higher) supports dynamic permission revocation at runtime [4].

Android kernel access control. The Linux kernel of Android provides many core services such as the process scheduling, sandbox isolation and inter-process communication, involving the UID-based and GID-based access control mechanisms. However, these kind of discretionary strategies will result in privilege escalation and permissions leakage. Since version 4.3, Android integrated the SELinux (Security-Enhanced Linux) and achieved mandatory access control based on the domain-type mechanism. Every process and file should be associated with a security context, and the defined policies in the system are fine-grained. When a process accesses target resource, all control conditions must be satisfied.

2.2. TrustZone isolation

ARM Trustzone is a hardware-based security technology, with the secure and normal world strictly separated. The software within the processor resides in only one of the isolated zone, and the secure resource cannot be accessed by normal world components. The TrustZone architecture sets a secure monitor mode and brings a rigorous mechanism into the secure world in order to control the context switch between two worlds, triggered by the SMC (Secure Monitor Call) instruction or other exceptions [5]. In general, ARM-Android takes advantage of the TrustZone isolation environment to robustly protect sensitive assets in the secure world, thus preventing possible attacks which are from the normal processor core.

2.3. B method

B method is a formal method based on the set theory and first order logic, and B development covers the entire system development cycle. There are two relevant activities: writing formal specifications and proving them. The first stage is to build abstract machines, which will be refined and finally transformed into fault-free concrete models. During modeling process, the second activity performs numerous type checks and theorem demonstrations to prove the correctness of specifications. After proved, the model can be coded in to C or Ada language. There are a number of mature tools supporting the B method,such as Atelier B and ProB [6].

3. Android permission model

3.1. Model structure

On the basis of realistic Android application framework, the model structure is shown in Fig.1. The permission and component machine are in the base layer, with the application system model above. Taking the ICC security into the consideration, we extract the necessary elements of entities and take them as the variables in different abstract machines.

With all static items elicited through the AndroidManifest.xml file, the status attribute of a component is defined to describe the communicating state dynamically, containing six constants: unavailable(un), free(fr), requiring(rq), waiting(wt), verifying(vf), calling(call), called(called). Therefore, the application invocation is regarded as the communication process with state migration under a control network. Assuming that all components in the system can only have one running instance, we use the 2-tuple (caller, callee) to indicate the status of communicating pairs, and then express the status transition process.

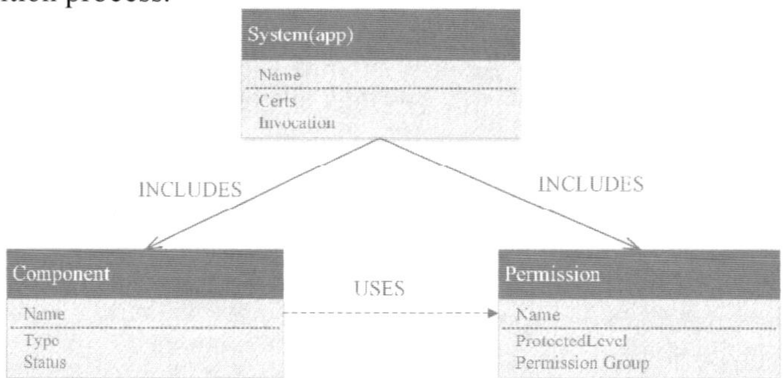

Figure 1. Structure of Android Permission formal model

3.2. Policy extraction

More than ten primary authorization policies in the Android middleware are obtained according to the new Android official document, and then reduced into six logical expressions [4]. For example, record the delegated permission set of an application A as ps, and A requests the permission P which belongs to the group gp. A is granted P if and only if

i) The user delegates the permissions in gp to A explicitly through the notification interface. or

ii) Some of permissions in gp are already contained in ps.

3.3. Formal permission framework

We establish the basic abstract machine *Permission*, *Component* and *Perm_Sys* based on the model structure in Fig.1, wherein the B method-based framework *Perm_Sys* includes *Component* and *Permission* and uses the variables and operations in them.

The overall operations in *Perm_Sys*, consist not only of manipulations on an application such as the installation, startup and permission assignment, but of the component communication simulation. There are ten crucial access-related operations defined in the formal model: four kinds of requests (Out_Req_1,In_Req_1,Out_Req_2, In_Req_2), a successful access (Succ_Access), the finish action (EndAccess), the dynamic permission revocation (Revoke_ac), the modification and restoration of component's status (Modify_ac, Restore_st, Clr_voidcall). These operations also concern the state transitions of a component. The state is migrated only if the invocation context complies with access control policy. Once the permission check fails, the component remains in its original state. Fig.2 presents the state transition process.

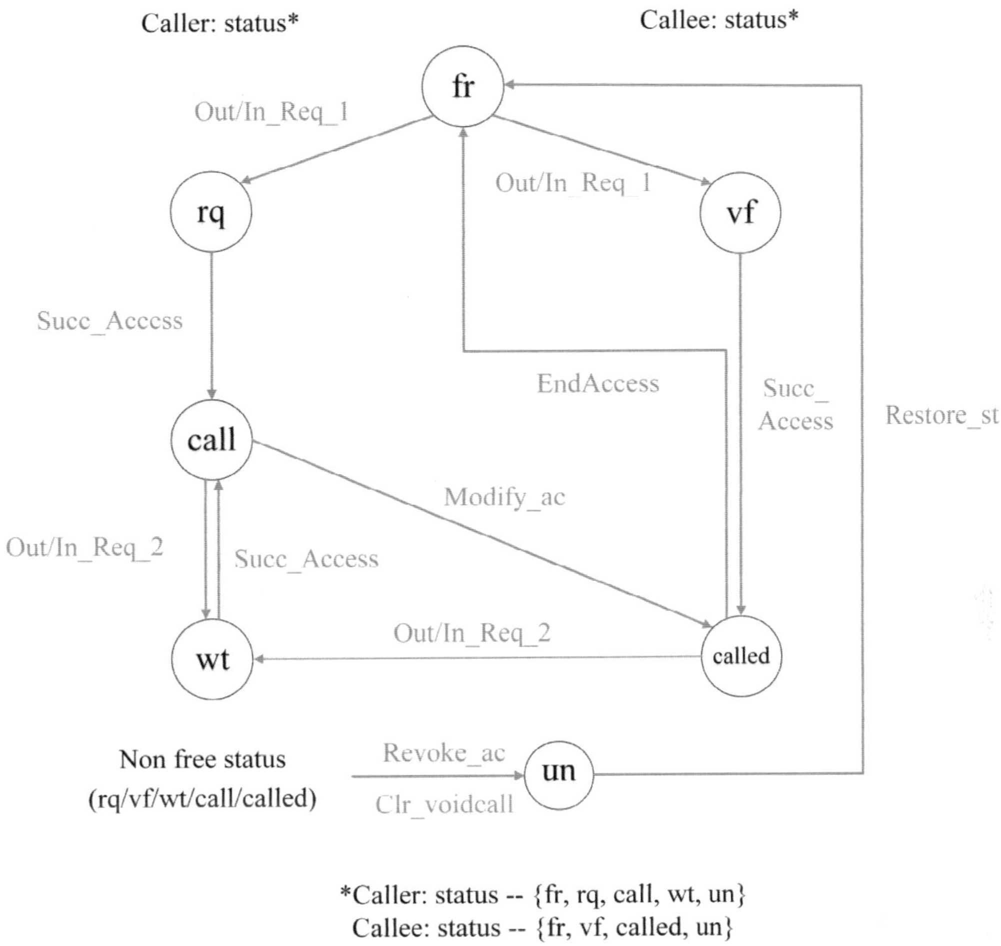

Figure 2. State transition of a component in *Perm_Sys*

4. PROCESS MODEL BASED ON ARM TRUSTZONE

4.1. Model elements

We abstract out three fundamental entities in the kernel and hardware layer to characterize access actions of a process. As highlighted in Fig.3, the process object consists of the identifier, security domain, execution and access status. The identifier represents the unique PID, which can be used in the traditional discretionary access control mechanism. The resource object is composed of one identifier, security domain, available state and buffer capacity. Specifically, the security domains introduced in the process and resource entity indicate their location in the hardware layer, either the secure world or normal world. Both the process and resource model need observe the rules defined in *AC_Policy*. The item is as fine-grained as the mandatory access control policy, only allowing the specific process performs whitelist operations on the certain resource.

The entire system machine is built on these basic models for accessing and communicating simulation of the process, including the process scheduling and hardware world switching, etc. There are four constants indicating realistic process execution states: ready, blocked, run and final. Each process has four access statuses: pre-access (pre), wait for accessing (wait), accessing (acing) and post-access (post).

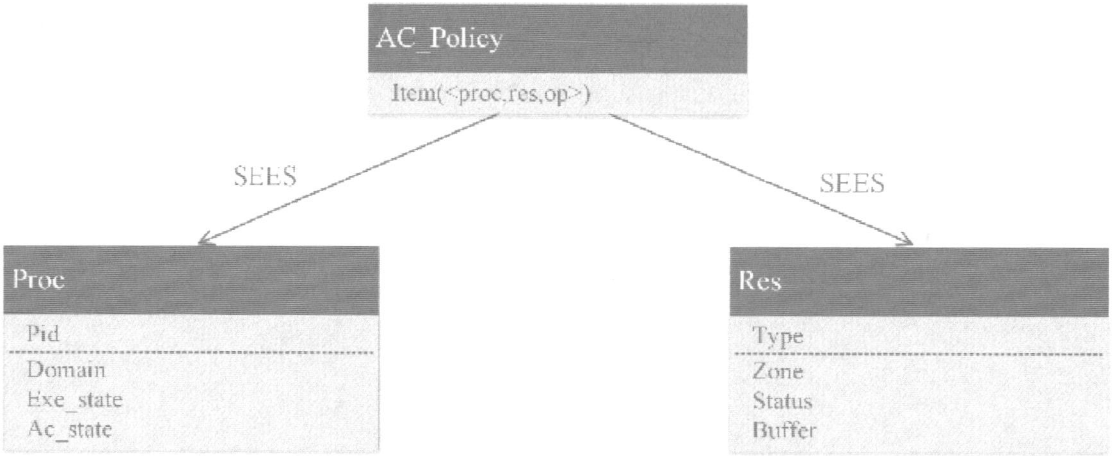

Figure 3. Structure of formal process model

4.2. Context switch in TrustZone

The access control on the TrustZone architecture largely relies on a secure context switch between two partitions under the monitor mode. Therefore, the access actions involving two different worlds and transition operations in the hardware layer need to be characterized. Concretely, the hardware domain attribute referring to current processor operating environment is set up, in order to ensure that it always match the security domain of the process and resource entity. Moreover, the process interrupt should be specified in the system.

According to the TrustZone whitepaper [7], we define eighteen operations for the context switch formalization. Take the hardware transition from the normal world to secure world for an instance. When the processor is executing in normal world, the dedicated instruction and exceptions cause it to switch into the monitor mode. The programs within monitor mode save the current world state and restore the status of secure world, that is, isolate the normal world and enable the hardware attribute as "secure" in the formal model, making all non-secure programs cannot access secure assets.

4.3. Formal process model with isolation

The formal process model details the resource access of a certain process in all possible situations, including the cross-world request, the secure context switch and so on. Because the process execution state is relevant to the access status, it is necessary to demonstrate the migration of both two states in the same access-related action. A process may have the following six central operations: the request (Pre_Access), a proper access (Pro_Access), waiting for occupied assets (Wait_Access), re-accessing the allowed resource (Re_Access), a failed access (Fail_Access) and the finish (End_Access). Fig.4 shows the primary state migration under operation conditions.

Generally, the access operations cover two-level control strategies: the security domain based and policy-based mechanism, corresponding to TrustZone hardware isolation and mandatory access control in the kernel respectively. When the context switch occurs in the hardware, the access process in switched world will be rebuilt.

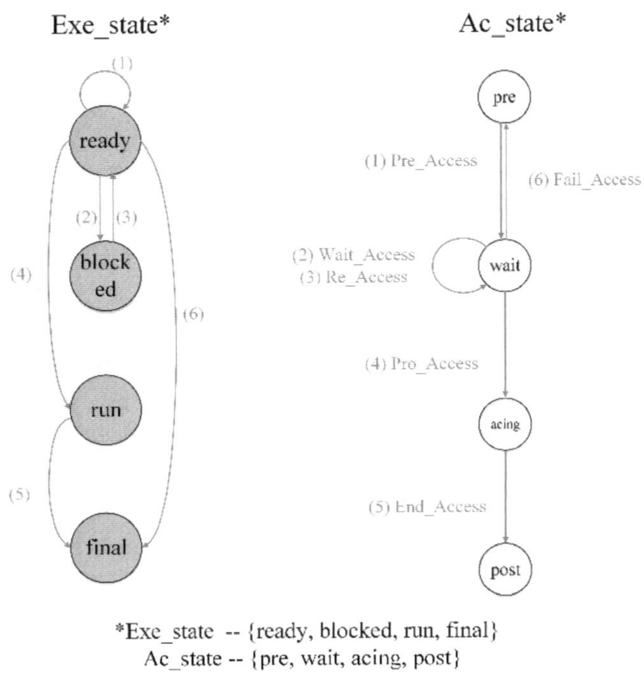

Figure 4. State transition of a process in *Pro_Sys*

5. MODEL EVALUATION

Our model development is based on the Atelier B4.3, all security properties translated into either invariants or operation conditions. Atelier checks the type of all model components and proves the specifications are well-defined. Table 1 and Table 2 summarize the size of formal Android permission framework and process model, with the number of POs (Proof Obligations) and the rate of automatic proofs.

Table 1. The number of POs in the permission model

Machine	Code Size	POs	Proved	Unproved	Auto-Proof Rate%
Permission	72	12	10	2	83%
Component	84	19	16	3	84%
Perm_Sys	245	147	91	56	62%

Table 2. The number of POs in the process model

Machine	Code Size	POs	Proved	Unproved	Auto-Proof Rate%
Proc	86	21	17	4	81%
Res	66	46	39	7	85%
Pro_Sys	320	473	362	111	77%

Typically, the high percentage of auto-proofs contributes to reduce the model complexity. The results show that our machines are well-formed with the correct grammar, and 75% of lemmas (535 out of 718) are demonstrated automatically, which is acceptable and rewarding. We are now working on proving remaining POs manually.

6. RELATED WORK

Formalization of Android permission framework was pioneered by Shin et al. [8] They utilized Coq to formalize the Android permission framework, and analyzed its safety and completeness. Similarly, Betarte et al. [9] enriched the model and formally proved some crucial properties. In the study of Fragkaki et al., [10] an abstract formal permission model was proposed with desired attributes. However, these model are too abstract to keep the practical solution in accordance with them completely. Since the B method is advantaged in developing the safety-critical software, our model can realize an executable module with consistency.

In terms of the trusted access control mechanism based on ARM TrustZone, Xu et al. [1] designed and realized a secure prototype system with multi policy, including the DTE (Domain and Type Enforcement) and BLP (Bell-La Padula) model. Kawamoritaet al. [11] developed a separation kernel for embedded system with two formal methods. They employed the B method to devise the key part, which guarantees the efficiency and high-assurance. Chang et al. [12] proposed a usage and access control model on the ARM-Android platform for permission leakage, rather than formal security analysis. Combined the Android software stack with ARM TrustZone, our model is for formalization of the mechanism specifications and invariant properties, providing an extensible framework for security enhancement.

7. CONCLUSION AND FUTURE WORK

For multi-tier access control mechanisms on the ARM-Android platform, this paper develops a formal model with the B method, including the Android permission model and process model with TrustZone isolation. Based on the Atelier B, different kinds of control policies are specified in the system. The evaluation results show that all descriptions of proposed models are type checked and 75% (535 out of 718) proof obligations are proved automatically. Generally, our formal framework is an attempt to design the access control module, and we are already trying to enriching and refining the model by adding more actual structures and actions. We are also interested in applying model checker to test the formal model, which is capable of implementing a practical control module consistent with formal specifications.

8. References

[1] Xu, Y., Pan, W., andZhang, X., 2008. Design and Implementation of Secure Embedded Systems Based on Trustzone. In International Conference on Embedded Software and Systems, 136-141. DOI= 10.1109/ICESS.2008.59.

[2] Kanonov, U. and Wool, A., 2016. Secure Containers in Android. Proceedings of the 6th Workshop on Security and Privacy in Smartphones and Mobile Devices - SPSM'16. DOI= 10.1145/2994459.2994470.

[3] Platform Architecture | Android Developers: 2017. https://developer.android.com/guide/platform/index.html. Accessed: 2017- 05- 01.

[4] Permission | Android Developers: 2017. https://developer.android.com/guide/topics/security/permissions.html. Accessed: 2017- 05- 01.

[5] Products | TrustZone – ARM: 2017. http://www.arm.com/products/security-on-arm/trustzone. Accessed: 2017- 05- 01.

[6] Presentation of the B method | Méthode B: 2017. http://www.methode-b.com/en/b-method/. Accessed: 2017- 05- 01.

[7] ARM, 2009. ARM Security Technology Building a Secure System using TrustZone Technology (white paper). ARM Limited.

[8] Shin, W., Kiyomoto, S., Fukushima, K., and Tanaka, T., 2010. A Formal Model to Analyze the Permission Authorization and Enforcement in the Android Framework. In *IEEE Second International Conference on Social Computing*, 944-951. DOI= 10.1109/SocialCom.2010.140.

[9] Betarte, G., Campo, J.D., Luna, C., and Romano, A., 2015. *Verifying Android's Permission Model*. Springer International Publishing. DOI= 10.1007/978-3-319-25150-9_28.
[10] Fragkaki, E., Bauer, L., Jia, L., and Swasey, D., 2012. Modeling and Enhancing Android's Permission System. Springer Berlin Heidelberg. DOI= 10.1007/978-3-642-33167-1_1.
[11] Kawamorita, K., Kasahara, R., Mochizuki, Y., and Noguchi, K., 2010. Application of Formal Methods for Designing a Separation Kernel for Embedded Systems. *World Academy of Science Engineering & Technology*, 68, 1303.
[12] Chang, R., Jiang L., Chen, W., He, H., and Yang, S., 2017. Enhancement of Permission Management for an ARM-Android Platform. In *IEEE International Conference on Computer and Information Technology*. DOI= 10.1109/CIT.2016.14.

Acknowledgments
Thanks to project supported by the National Natural Science Foundation of China (No. 61572516).

Enhancement Approachof Object Constraint Language Generation

Samin Salemi and Ali Selamat

Faculty of Computing, Universiti Teknologi Malaysia, 81310 UTM Skudai, Johor, Malaysia

saminsalemi127@gmail.com

Abstract.OCL is the most prevalent language to document system constraints that are annotated in UML. Writing OCL specifications is not an easy task due to the complexity of the OCL syntax. Therefore, an approach to help and assist developers to write OCL specifications is needed. There are two approaches to do so: First, creating an OCL specifications by a tool called COPACABANA. Second, an MDA-based approach to help developers in writing OCL specification by another tool called NL2OCLviaSBVR that generates OCL specification automatically. This study presents another MDA-based approach called En2OCL, and its objective is twofold. 1- to improve the precison of the existing works. 2- to present a benchmark of these approaches. The benchmark shows that the accuracy of COPACABANA, NL2OCLviaSBVR, and En2OCL are 69.23, 84.64, and 88.40respectively.

1. Introduction

An OCL specification, is a Boolean expression that sets a condition on an entity, such as a class, attribute, data-type, and operation in UML models. It means that the condition must be true for all instances of a given entity. An OCL specification includes two main parts: context and expression body. The context of a OCL specification presents the entity restricted by the OCL specification, and the expression body of an OCL specification displays a Boolean condition. The entity, which is restricted by an OCL specification, is identified as a context variable of the OCL specification. Figure 1 presents the template of an OCL specification.

```
Context [contextVariable] [stereotype]:
[ExpressionBody]
```

Figure 1. OCL specification template

The complexity and difficulty of the OCL syntax causes some effects such as rising time and effort needed to create OCL specifications, and occurring errors while writing OCL specifications. The problems are motivations for embarking on this research. The main goal of the current research is to simplify the design phase of software modeling by proposing an MDA-based approach, to generate OCL specifications automatically. The existing tools for generating OCL specifications, use pattern-based and MDA-based approaches. The pattern-based approach can analyze the consistency of the constraint specifications automatically. This approach uses libraries to write OCL specifications. The existing tool implementing the pattern-based approach is COPACABANA, which semi-automatically generates OCL specifications. The existing tool implementing the MDA-based approach is NL2OCLviaSBVR, which automatically generates OCL specifications by using Sitra.

2. Metamodels

Metamodel is the abstract syntax of a modeling language expressed as a model. The metamodel defines the structure of the model in terms of classes and relationships.

2.1. English Metamodel

In English metamodel, prefix element is a semantic element placed immediately before another semantic element. For example, in "valid customer card" sub-phrase, "valid" is a prefix element, because "valid" is an attribute and "customer card" is a class. Universal quantifiers are "all", "each", and "every". Existential quantifiers are "a", "an", "any", "s (plural), and "the". Negation element is "not". IsPropertyOfSign is a sub-phrase that links two semantic elements using "of". For example, "age of customer" is an IsPropertyOfSign. Valued range quantifier is a sub-phrase that determines a quantity by this syntax: "between quantity1 and quantity2" such as "between 3 and 5" and "between the number of customer cards and their owners". Possessive determiner is a sub-phrase of determiners that modify a noun by attributing possession to someone or something such as "customer's card". SignIntegratedWithAnd is a sub-phrase presenting a multiplication, division, addition, or subtraction of two quantitative things. For example, "the multiplication of the customer's age and the service's point" is a SignIntegratedWithAnd phrase. SumOf is a SignIntegratedWithAnd that adds two quantitative things. SubtractionOf is a SignIntegratedWithAnd that subtracts two quantitative things.MultiplicationOf is a SignIntegratedWithAnd that multiplies two quantitative things. DivisionOf is a SignIntegratedWithAnd that divides two quantitative things. Necessity verb is a modal verb showing a necessary action that must be performed. Transitive verb is a verb that takes one or more objects. Copular verb is a verb that links a subject to a complement that refers to the subject. Preposition conjunction is a preposition, such as "in" and "with", describing a relationship between two sub-phrases in a sentence. Str is a string. IsEqualTo can be "is", "are", "equals to", "equal to", "is equal to", or "are equal to".

2.2. SBVR Metamodel

SBVR is a standard to develop semantic models of business vocabularies and business rules [1]. SBVR concepts are object type and fact type. Object type is a general concept that is classified based on its characteristics. Fact type identifies a relationship among one or more object type. The object type in a fact type is called fact type role. Unary fact type (characteristic) has one fact type role, and binary fact type has two fact type roles. SBVR logical formulations are modal formulations, atomic formulation, logical operation, quantification, and objectification. Modal formulations used to formulate modality are divided into necessity and possibility. Necessity formulation is represented using the keywords "It is necessary" or "It is obligatory". Possibility formulation is represented using "It is possible" keyword. Atomic formulation specifies a fact type in a rule. Binary atomic formulation specifies a binary fact type in a rule. Quantification is a set of quantifications supported in SBVR and consists of universal quantification, at least n quantification, at most n quantification, etc. Objectification is a logical formulation that involves a bindable target. Logical operations are divided into logical negation and binary logical negation. Logical negation is a logical operation having one logical operand. Binary logical negation, such as conjunction, disjunction, and implication, is a logical operation having two logical operands.

2.3. OCL Metamodel

The context of an OCL specification is specified in the expression body using the self-keyword. An OCL context has a stereotype that can be one of these items: invariant, definition, initial, derivation, pre and post-condition, and body. An OCL invariant expression specifies conditions on attributes and operations of a class in a Class diagram in form of arithmetic and logical operations. An attribute or association can be added to a Class diagram by OCL definitions. The initial value of attributes or associations can be specified by OCL initials. A derived value of an attribute or association end can be indicated by an OCL derivation. A pre-condition expression for an operation must be true before the operation execution and a post-condition expression for an operation must be true after the operation execution. OCL bodies are used to express query operation. The expression body of an OCL specification includes one or more of expressions such as let, if, literal, call, and unary operation. A let expression can define a new variable that has an initial value. An *if* expression defines a condition and two alternative expressions that after evaluating the condition, one of the alternative expressions is selected as the result of the if expression. A literal expression does not have any argument to produce a value. This kind of expression results in an expression symbol, such as an integer (12) and a string ("hello"). A call expression refers to an attribute, an operation, or an iterator for a collection, which is a collection of some values. An attribute call expression is a reference to a class's attribute in a UML model. A navigation call expression is a reference to a relationship in a UML model. An operation call expression is used, when we want to refer to an operation of a classifier. There are some unary operations, such as not Empty, is Empty, oclIsTypeOf, to perform functions on a value. The not Empty operation on a collection returns true when the collection has at least one element. The is Empty on a collection returns true when the collection has no element. The oclIsTypeOf operation on a value determines if a value is of the type given to the operation as a parameter. When we want to construct a loop over a collection, a loop expression is used. The for All operation takes an expression as a parameter and results to true if the expression is evaluated to true for all elements in the collection. The exists operation on a collection specifies a Boolean expression that must be true for at least one element of the collection. The select operation on a set, takes a parameter expression, and resulting a sub-set of the set, when the parameter expression is true for all elements of the resulting sub-set. The reject operation on a set takes a parameter expression and results to a sub-set of the set, when the parameter expression is false for all elements of the resulting sub-set. The collect operation on a collection gives the set of all values for a certain attribute of all objects in the collection.

3. Royal and Loyal Model

Warmer and Kleppe (1999) originally introduced the Royal & Loyal model [2]. Afterward, the model was used in various publications [3, 4, 5]. "*Royal and Loyal (R&L) models the computer system of a fictional company. It handles loyalty programs for companies that offer their customers various kinds of bonuses. Often, the extras take the form of bonus points or air miles, but other bonuses are possible as well: reduced rates, a larger rental car for the same price as a standard rental car, extra or better service on an airline, and so on. Anything a company is willing to offer can be a service rendered in a loyalty program*" [6]. Benchmark test cases which used in this research have been extracted from the Royal and Loyal model illustrated in Figure 2.

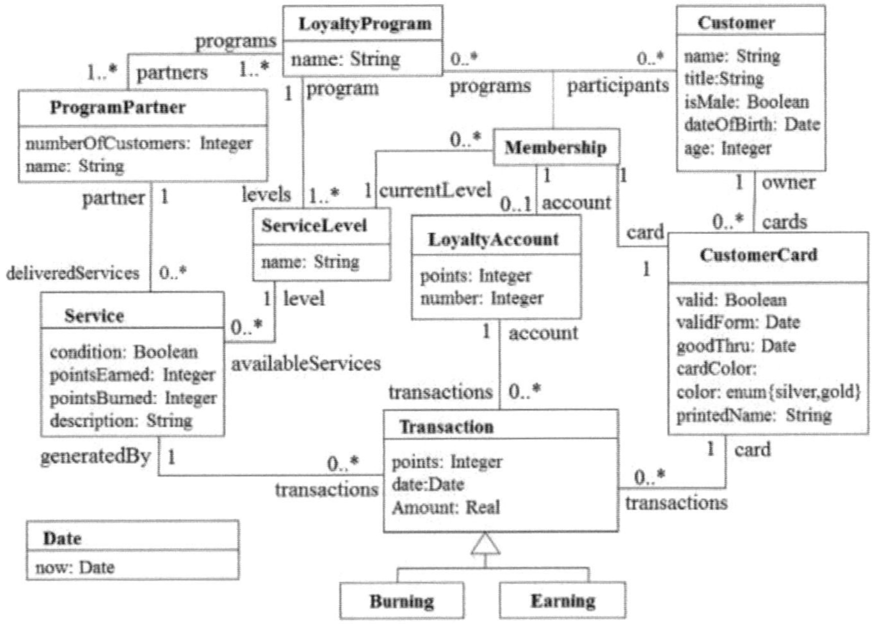

Figure 2. Royal and Loyal model [7]

4. En2OCL approach

The En2OCL approach proposed in this research, is based on the MDA approach to automatically transform system constraints formed in English sentences into OCL specifications. The proposed approach takes two inputs involving: an UML Class model, which presents a system and an English sentence, which presents a constraint of the system. The output of the proposed approach is a specification, which presents the system constraint in form of the OCL syntax. As Figure 3 illustrates, the approach contains three major analyses involving: lexical, syntactical, and semantic analysis involves.

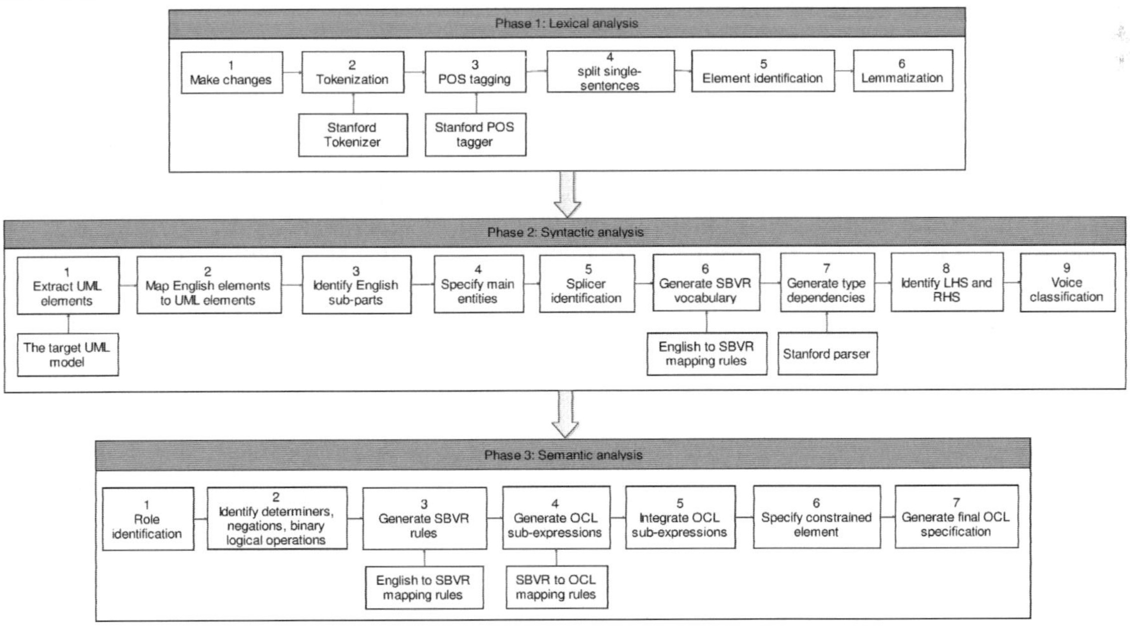

Figure 3. EN2OCL approach

The proposed approach, uses SBVR to bridge English elements to OCL elements. The Input English sentence is analyzed to extract business vocabulary and rules. The business vocabulary and rules are transformed into OCL specifications. Thus, two sets of mapping rules are created. The mapping rules that constitute the transformation are summarized in Table 1.

4.1. Lexical analysis

The lexical analysis includes six steps. In the first step, some parts of the input English sentence are changed. For example, "not more than" is changed to "at most". In the two next steps, the changed sentence is tokenized and tagged by the Stanford Tokenizer and POS tagger. In the fourth step, the English sentence is split into single-sentences using the POS tags. In addition, duplicate single-sentences and single-sentences from which no business rules can be extracted are removed. In the fifth steps, nouns, adjectives, strings, and numeral elements are extracted from each single-sentence. In the sixth step, the extracted elements are lemmatized.

4.2. Syntactical analysis

The syntactical analysis of the input English sentence contains nine steps. In the first step, the elements of the UML Class model, such as classes, attributes, ends, and data-type are extracted. In the second step, the nouns and adjectives extracted from the English sentence are mapped to the elements extracted from the UML Class model. Strings, and numeral elements extracted from the lexical analysis, and then mapped elements are saved in an array. In the third step, English sub-parts, such as possessive determiners and prefix element, are identified. In the fourth step, main entities are selected. The main entities should not include data-type, possessive determiner, and prefix element. In the fifth step, the splicer between the main elements is specified. In the sixth step, the mapping rules are used to generate the SBVR vocabulary from the elements saved in the array and the splicers. In the seventh step, the type dependencies between the extracted elements are identified using the Stanford typed dependency parser. In the eighth step, the Left Hand Side (LHS), and Right Hand Side (RHS) elements are identified using the type dependencies. Left/right hand side element is the element placed in the left/right hand side of a verb or preposition conjunction. In the ninth step, it is determined that the splicer is active or passive. If the splicer is an active verb, the splicer has an active voice. However, if the splicer is a passive verb, the splicer has a passive voice.

4.3. Semantic analysis

The semantic analysis phase contains seven steps. In the first step, the role of characteristic fact type is specified. In addition, this step specifies the state of each binary fact type. In the second step, determiners, negation elements, and binary logical elements are extracted from each of the single-sentences. In the third step, SBVR rules are generated using the mapping rules. In this step, business rules are generated for the semantic formulations. In the fourth step, the business vocabulary and rules are translated to OCL sub-expressions using the mapping rules. In the fifth step, the sub-expressions are integrated. The sixth step determines which class of the UML Class model is being constrained. This Class is specified as the constrained element. In the seventh step, the final OCL specification is generated and revised.

Table 1. Mapping rules

Rule	English element	SBVR element	OCL element
Rule1	Necessity verb	Necessity	Constraint
Rule2	IsPropertyOfSign	Atomic formulation	AttributeCallExp
			NavigationCallExp
			Collect
Rule3	Possessive determiner	Atomic formulation	AttributeCallExp
			NavigationCallExp
			Collect
Rule4	Prefix element	Atomic formulation	AttributeCallExp
			Select
Rule5	Prefix element	Equivalence	EqualBool
			Select
Rule6	Transitive verb/ Copular verb	Atomic formulation	AttributeCallExp
			Collect
Rule7	Preposition conjunction	Atomic formulation	AttributeCallExp
			Collect
Rule8	Transitive verb/ Copular verb/ Preposition conjunction	Binary atomic formulation	Includes
			NavigationCallExp
			Collect
Rule9	Numeral	Number	NumericLiteralExp
Rule10	Noun (mapped to enumElement/enumName)	Objectification	Variable
Rule11	Noun (mapped to attribute/end)	Atomic Formulation	AttributeCallExp
			NavigationCallExp
Rule12	Noun (mapped to Class)	Object type	UMLElement(Class)
			OclIsTypeOf
			Select
			Size
Rule13	SignIntegratedWithAnd	Quantity	SignOpr
Rule14	SumOf	Quantity	Addition
Rule15	SubtractionOf	Quantity	Subtraction
Rule16	MultiplicationOf	Quantity	Multiply
Rule17	DivisionOf	Quantity	Division
Rule18	Sign	Quantity	RelationalOpr
Rule19	IsLessThan	Quantity	Less
Rule20	IsMoreThan	Quantity	More
Rule21	IsAtLeast	Quantity	AtLeast
Rule22	IsAtMost	Quantity	AtMost
Rule23	Str	Text	StringLiteralExp
Rule24	Existential quantifier (conditional)	Existential	notEmpty
Rule25	Universal quantifier	Universal	ForAll
Rule26	IsOrEqualOrEquals	Quantity	EqualNum
		Equivalence	EqualBool
		Equivalence	Select
		Quantity	Select
Rule27	AtMostN	AtMostN	AtMost
Rule28	AtMostOne	AtMostOne	AtMost
Rule29	AtLeastN	AtLeastN	AtLeast
Rule30	AtLeastOne	Existential	notEmpty
Rule31	MoreThanN	Quantity	More
Rule32	LessThanN	Quantity	Less
Rule33	ExactlyN	ExactlyN	EqualNum
Rule34	ExactlyOne	ExactlyOne	EqualNum

Rule35	Valued range quantifier	Numeric range	And
Rule36	Negation element	Logical negation	Not
			IsEmpty
			Excludes
Rule37	Binary logical operation	Binary logical Operation	LogicalOpr
Rule38	And	Conjunction	And
Rule39	Or	Disjunction	Or
Rule40	Conditional	Implication	Implies

5. Accuracy Benchmark Results

There are twenty six general test cases, which are used for benchmark of OCL generator tools. These twenty six benchmark test cases extracted from the Royal and Loyal case study have been applied on the three existing tools. The test cases and outputs of the three tools are presented below:

Constraint 1
English: *Every customer who enters a loyalty program must be of legal age.*
COPACABANA:context Customer inv legalAge: AttributeValueRestriction (Customer,age,>=,18)
NL2OCLviaSBVR: context Customer inv self.age >=
En2OCL: context loyaltyprogram inv: self.participants

In Constraint 1, the output of COPACABANA is complete and the outputs of NL2OCLviaSBVR and En2OCL are incomplete.
Constraint 2
English: *Male customers must be approached using the title "Mr".*
COPACABANA:-
NL2OCLviaSBVR: context Customer inv self.isMale implies self.title= Mr.
En2OCL: context customer inv: self->select(s|s.gender::male).title->exists(e|e="mr")

In Constraint 2, the output of COPACABANA is missed and the outputs of NL2OCLviaSBVR and En2OCL are complete.
Constraint 3
English: *The number of valid cards for every customer must be equal to the number of programs in which the customer participates.*
COPACABANA:context Customer inv maleTitle: IfThenElse ({ AttributeValueRestriction (Customer,isMale,=,true)}, AttributeValueRestriction (Customer,title ,=, 'Mr.'),)
NL2OCLviaSBVR: context Customer inv self.cards->select(valid=true)->size() =self.programs ->size()
En2OCL: context customer inv:self->forAll(a|a.programs->size()=a.cards-> select(s|s.valid)->size())

In Constraint 3, the outputs of COPACABANA, NL2OCLviaSBVR, and En2OCL are complete.
Constraint 4
English: *The validFrom date of customer cards should be earlier than goodThru.*
COPACABANA:-
NL2OCLviaSBVR: context CustomerCard inv self.validFrom <self.goodThru
En2OCL: context customercard inv: self.validFrom <self.goodThru

---

In Constraint 4, the output of COPACABANA is missed and the outputs of NL2OCLviaSBVR and En2OCL are complete.
Constraint 5
English: *The birthdate of the owner of a customer card must not be in the future.*
COPACABANA:-

IOP Conf. Series: Journal of Physics: Conf. Series **933** (2017) 012008 doi:10.1088/1742-6596/933/1/012008

NL2OCLviaSBVR: context CustomerCard inv self.owner.dateOfBirth <> future
En2OCL: context customercard inv: self.owner.dateOfBirth.isBefore(Date::now)

In Constraint 5, the output of COPACABANA is missed and the outputs of NL2OCLviaSBVR and En2OCL are incomplete.

Constraint 6
English: *The owner of a customer card must participate in at least one loyalty program.*
COPACABANA:context CustomerCard inv programParticipation: MultiplicityRestriction (CustomerCard,owner.programs,>,0)
NL2OCLviaSBVR: context CustomerCard inv self.owner.programs -> Size()>= 1
En2OCL: context customercard inv: self.owner.programs->notEmpty()

In Constraint 6, the outputs of COPACABANA, NL2OCLviaSBVR, and En2OCL are complete.

Constraint 7
English: *There must be at least one transaction for a customer card with at least 100 points.*
COPACABANA:context CustomerCard inv transactionPoints: Exists (transactions, {AttributeValueRestriction (CustomerCard,points,>,100)}
NL2OCLviaSBVR: context CustomerCard inv self.transaction->select(point >= 100)-> Size()>= 1
En2OCL: context customercard inv: self.transactions->exists(e|e.points>=100)

In Constraint 7, the outputs of COPACABANA, NL2OCLviaSBVR, and En2OCL are complete.

Constraint 8
English: *The service level of each membership must be a service level known to the loyalty program.*
COPACABANA:context Membership inv knownServiceLevel: ObjectInCollection (Membership, programs.levels, currentLevel)
NL2OCLviaSBVR: context Membership inv self.currentLevel ->includes (self.program.levels)
En2OCL: context loyaltyprogram inv: self.memberships->forAll (a|a.currentLevel =self.currentLevel)

In Constraint 8, the outputs of COPACABANA, NL2OCLviaSBVR, and En2OCL are complete.

Constraint 9
English: *The participants of a membership must have the correct card belonging to this membership.*
COPACABANA:context Membership inv correctCard: bjectInCollection(Membership, participants .cards, card)
NL2OCLviaSBVR: Context Membership inv self.participants.cards ->includes(self.card)
En2OCL: context membership inv: self.participants.cards ->includes(self.card)

In Constraint 9, the outputs of COPACABANA, NL2OCLviaSBVR, and En2OCL are complete.

Constraint 10
English: *The color of a membership's card must match the service level of the membership.*
COPACABANA:context Membership inv levelAndColor1: IfThenElse({AttributeValueRestriction (Membership, currentLevel.name,=,'Silver')}, AttributeValueRestriction (Membership,card.color,=,Color::silver),) inv levelAndColor2: IfThenElse({AttributeValueRestriction (Membership, currentLevel.name,=,'Gold')}, AttributeValueRestriction (Membership,card.color,=,Color::gold),)
NL2OCLviaSBVR: context Membership inv self.card.color = self.currentLevel.name
En2OCL: context membership inv: self.currentlevel.name = "silver" implies card.cardcolor = color::silver and self.currentLevel.name = "gold" implies card.cardcolor = color::gold

---

In Constraint 10, the outputs of COPACABANA, NL2OCLviaSBVR, and En2OCL are complete.

IOP Conf. Series: Journal of Physics: Conf. Series **933** (2017) 012008 doi:10.1088/1742-6596/933/1/012008

Constraint 11
English: *Memberships must not have associated accounts.*
COPACABANA:context Membership inv noAccount: MultiplicityRestriction (Membership,account,=,0)
NL2OCLviaSBVR: context Membership inv self.account -> isEmpty()
En2OCL: context Membership inv self.account -> isEmpty()

In Constraint 11, the outputs of COPACABANA, NL2OCLviaSBVR, and En2OCL are complete.
Constraint 12
English: *Loyalty programs must offer at least one service to their customers.*
COPACABANA:context LoyaltyProgram inv minServices: MultiplicityRestriction (LoyaltyProgram,partners.deliveredServices,>=,1)
NL2OCLviaSBVR: context LoyaltyProgram inv self.partners.deliveredServices->size() >= 1
En2OCL: context loyaltyprogram inv: self.partners.deliveredservices->notEmpty()

In Constraint 12, the outputs of COPACABANA, NL2OCLviaSBVR, and En2OCL are complete.
Constraint 13
English: *If none of the services offered in a loyalty program credits or debits the loyalty accounts, then these instances are useless and should not be present.*
COPACABANA:-
NL2OCLviaSBVR: context LoyaltyProgram inv
En2OCL: -

In Constraint 13, the outputs of COPACABANA and En2OCL are missed and the output of NL2OCLviaSBVR is incomplete
Constraint 14
English: *The name of the first level must be "Silver".*
COPACABANA:-
NL2OCLviaSBVR: context LoyaltyProgram invself.levels->first().name = Silver
En2OCL: context loyaltyprogram inv: self.levels->first ().name = “silver”

In Constraint 14, the output of COPACABANA is missed and the outputs of NL2OCLviaSBVR and En2OCL are complete
Constraint 15
English: *There must exist at least one service level with the “basic” name.*
COPACABANA:context LoyaltyProgram inv basicLevel: Exists(levels ,{ AttributeValueRestriction (LoyaltyProgram,name,=,’basic’)})
NL2OCLviaSBVR: context ServiceLevel inv self.name = basic->exists(()
En2OCL: context servicelevel inv: self->exists(e|e.name->exists(e|e="basic"))

In Constraint 15, the outputs of COPACABANA and En2OCL are complete and the output of NL2OCLviaSBVR is incomplete
Constraint 16
English: *The number of participants in a loyalty program must be less than 10,000.*
COPACABANA:context LoyaltyProgram inv maxParticipants: MultiplicityRestriction (LoyaltyProgram,participants,<,10000)
NL2OCLviaSBVR: context LoyaltyProgram inv: self.participants->size()<10000
En2OCL: context loyaltyprogram inv: self.participants->size()<10000

In Constraint 16, the outputs of COPACABANA, NL2OCLviaSBVR, and En2OCL are complete.

IOP Conf. Series: Journal of Physics: Conf. Series **933** (2017) 012008 doi:10.1088/1742-6596/933/1/012008

Constraint 17
English: *The number of the loyalty account must be unique within a loyalty program.*
COPACABANA:-
NL2OCLviaSBVR: context LoyaltyProgram inv self.Membership.account->isUnique(acc|acc.numbers)
En2OCL: context loyaltyprogram inv: self.membership.account->size()=1

In Constraint 17, the output of COPACABANA is missed and the outputs of NL2OCLviaSBVR and En2OCL are complete.
Constraint 18
English: *The names of all customers of a loyalty program must be different.*
COPACABANA:context LoyaltyProgram inv uniqueNames:
UniqueSetIdentifier(LoyaltyProgram,participants,name)
NL2OCLviaSBVR: context LoyaltyProgram inv self.participants.name->forAll(c1,c2|c1.name<>c2.name)
En2OCL: context loyaltyprogram inv: self.participants.name->forAll(c1,c2|c1.name<>c2.name)

In Constraint 18, the outputs of COPACABANA, NL2OCLviaSBVR, and En2OCL are complete.
Constraint 19
English: *The maximum age of participants in loyalty programs is 70.*
COPACABANA:context LoyaltyProgram inv maxAge: ForAll(participants,
{AttributeValueRestriction (LoyaltyProgram,age,<=,70)})
NL2OCLviaSBVR: context LoyaltyProgram inv: self.participants -> forAll(age<= 70)
En2OCL: context loyaltyprogram inv: self.participants.age<70

In Constraint 19, the outputs of COPACABANA, NL2OCLviaSBVR, and En2OCL are complete.
Constraint 20
English: *There may be only one loyalty account that has a number lower than 10,000.*
COPACABANA:-
NL2OCLviaSBVR: context LoyaltyProgram inv self.Membership.account->one(number < 10,000)
En2OCL: context loyaltyaccount inv: self->one(o|o.number < 10000)

In Constraint 20, the output of COPACABANA is missed and the outputs of NL2OCLviaSBVR and En2OCL are complete.
Constraint 21
English: *The attribute numberOfCustomers of class programPartner must be equal to the number of customers who participate in one or more loyalty programs offered by this programPartner.*
COPACABANA:context ProgramPartner inv nrOfParticipants: MultiplicityRestriction
(ProgramPartner,programs.participants,=,numberOfCustomers)
NL2OCLviaSBVR: context ProgramPartner inv self.numberOfCustomers= programs.participants->asSet()->size()
En2OCL: context programpartner inv: self.numberofcustomers= self.programs.participants->size()

In Constraint 21, the outputs of COPACABANA, NL2OCLviaSBVR, and En2OCL are complete.
Constraint 22
English: *A maximum of 10,000 pointsEarned may be earned using services of one partner.*
COPACABANA:-
NL2OCLviaSBVR: context ProgramPartner inv self.deliveredServices. pointsEarned<=10,000
En2OCL: context partner inv: self.deliveredservices.pointsEarned <=10000

In Constraint 22, the output of COPACABANA is missed and the outputs of NL2OCLviaSBVR and En2OCL are complete.

Constraint 23

English: *All cards that generate transactions on the loyalty account must have the same owner.*

COPACABANA:context LoyaltyAccount inv oneOwner: MultiplicityRestriction (LoyaltyAccount,transactions.card.owner,=,1)

NL2OCLviaSBVR: context LoyaltyAccount inv self.transactions.cards.owner->asSet()->size() = 1

En2OCL: context loyaltyaccount inv: self.transactions.cards->forAll(a1,a2|a1.owner=a2.owner)

In Constraint 23, the outputs of COPACABANA, NL2OCLviaSBVR, and En2OCL are complete.

Constraint 24

English: *If the points in a loyalty account is greater than zero, there exists a transaction with more than zero points.*

COPACABANA:context LoyaltyAccountinv positivePoints: IfThenElse({ AttributeValueRestriction (points,>,0)}, Exists(transactions,{ AttributeValueRestriction (LoyaltyAccount,points,>,0)}),)

NL2OCLviaSBVR: context LoyaltyAccount inv if (self.points > 0) then transaction -> exists(t| t.points>0) endif

En2OCL: context loyaltyaccount inv: self.points>0 implies self.transaction->exists(e|e.points>0)

In Constraint 24, the outputs of COPACABANA, NL2OCLviaSBVR, and En2OCL are complete.

Constraint 25

English: *There must be one transaction with exactly 500 points.*

COPACABANA:context LoyaltyAccount inv 500points: Exists(LoyaltyAccount, transaction.points,{ LiteralOcl (–,"self = 500")})

NL2OCLviaSBVR: context Transaction inv self.transaction->select(point = 500)->Size()=1

En2OCL: context transaction inv: self->one(o|o.points=500)

In Constraint 25, the outputs of COPACABANA, NL2OCLviaSBVR, and En2OCL are complete.

Constraint 26

English: *The available services for a service level must be offered by a partner of the loyalty program to which the service level belongs.*

COPACABANA:context ProgramPartner inv servicePartner: ObjectInCollection (ProgramPartner,delivered Services.level .program.partners,Sequence{})

NL2OCLviaSBVR: context ServiceLevel inv self.program.partners-> includesAll(self.availableServices.partner)

En2OCL: context servicelevel inv: self.program.partners->includesAll (self.availableservices.partner)

In Constraint 26, the outputs of COPACABANA, NL2OCLviaSBVR, and En2OCL are complete.

COPACABANA, NL2OCLviaSBVR, and En2OCL have resulted in 18, 22, and 23 correct outputs, respectively. Thus, En2OCL had 19.23% accuracy improvement in comparison with COPACABANA and 3.84% accuracy improvement in comparison with NL2OCLviaSBVR. Figure 4 exhibits the accuracy of the tools.

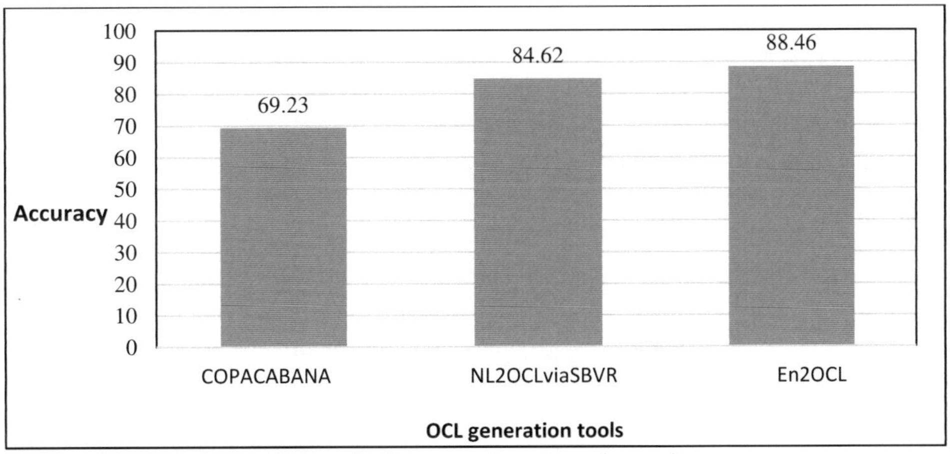

Figure 4. Accuracy benchmark results

6. Conclusion

The En2OCL approach proposed in this study is a transformation model to transform system constrains formed in English sentences into OCL specifications. The proposed approach overcomes English ambiguities using SBVR as an intermediate representation between English sentences and OCL specifications. En2OCL includes 21 major steps for generating an OCL specification from an English sentence. the proposed approach called En2OCL has been compared with the existing works (COPACABANA and NL2OCLviaSBVR). The accuracy comparison shows that the accuracy of En2OCL is 19.23% more than COPACABANA and 3.84% more than NL2OCLviaSBVR.

7. Reference

[1] OMG. (2013). Semantics of Business Vocabulary and Business Rules (SBVR), v1.2.

[2] Warmer, J., & Kleppe, A. (1999). Object Constraint Language: Precise Modeling with UML. Addison Wesley.

[3] Wahler, M. (2008). Using Patterns to Develop Consistent Design Constraints. ETH Zurich, Switzerland.

[4] Wilke, C., Thiele, M., & Wende, C. (2010). Extending Variability for OCL Interpretation. 13th International Conference on Model Driven Engineering Languages and Systems (MODELS 2010). October 3-8. Oslo, Norway.

[5] Bajwa, I. S., M. Lee, et al. (2012). "Translating natural language constraints to OCL." King Saud University–Computer and Information Sciences.

[6] Warmer, J. and A. Kleppe (2003). The Object Constraint Language: Getting Your Models Ready for MDA, Addison Wesley.

[7] Sharma, M., & Vishwakarma, R. G. Formalization & data abstraction during use case modeling in object oriented analysis & design. 3th International Conference on Computer Science, Engineering & Applications (ICCSEA 2013). May 24-26. Delhi, India: 67–75.

Research on Crowdsourcing Emergency Information Extraction of Based on Events' Frame

Bo Yang[1], Jizhou Wang[2], Weijun Ma[2], Xi Mao[2]

[1] School of Mapping and Geographical Science, Liaoning Technical University, Fuxin, Liaoning, China, 123000.
[2] Chinese Academy of Surveying & Mapping, Beijing, China, 100830.

Corresponding author. Tel.: +86 15810839035; email: michaelyangbo@outlook.com

Abstract. At present, the common information extraction method cannot extract the structured emergency event information accurately; the general information retrieval tool cannot completely identify the emergency geographic information; these ways also do not have an accurate assessment of these results of distilling. So, this paper proposes an emergency information collection technology based on event framework. This technique is to solve the problem of emergency information picking. It mainly includes emergency information extraction model (EIEM), complete address recognition method (CARM) and the accuracy evaluation model of emergency information (AEMEI). EIEM can be structured to extract emergency information and complements the lack of network data acquisition in emergency mapping. CARM uses a hierarchical model and the shortest path algorithm and allows the toponomy pieces to be joined as a full address. AEMEI analyzes the results of the emergency event and summarizes the advantages and disadvantages of the event framework. Experiments show that event frame technology can solve the problem of emergency information drawing and provides reference cases for other applications. When the emergency disaster is about to occur, the relevant departments query emergency's data that has occurred in the past. They can make arrangements ahead of schedule which defense and reducing disaster. The technology decreases the number of casualties and property damage in the country and world. This is of great significance to the state and society.

1. Introduction

Nowadays, there are various natural and social emergencies in the world, which has caused serious losses to the people and the country. When an emergency occurs, the aid department need the mapping department to provide quickly emergency mapping data [1]. With the development of the society, the traditional mapping which focuses on the acquisition of basic geographic information cannot meet the needs of emergency mapping timely, efficient and low cost. However, we found that the multisource data contain a large number of thematic geographic information and can solve the problem which emergency mapping can promptly access information. If you can extract information from the massive source of data to extract the feature geographic information about emergency mapping, this will extremely enrich the content of emergency geographic information and improve an emergency support capability.

Currently, these main methods of information extraction include semi-structured approach, plain text way, and common tools which provide a stable technical and theoretical basis for general information abstract [2]. Yet, these methods are only appropriate for it which includes a generic scene, a specific language, and the specified text type. Meanwhile, these measures cannot extract emergency geographic information, recognize the full address, and analyze the extractability results. Emergency geographic information has become the research hot spot of geographic information science, which different experts have different research aspects. Reference [3] presents a model and application of information extractibility based on the ontology of emergency cases. First, the model uses the ontology to describe the case in the field of emergency knowledge. Then, the model is applied to draw cases under the guidance of domain knowledge and build case libraries. Finally, the case library is provided to emergency rescue decision makers. Reference [4] proposes a method of information extraction based on rule matching for emergency events which match the text according to the rules that match the incident and is used to normalize the extraction results. Experiments show that the method is very effective. Reference [5] raises a technique based on the extraction and rapid release of critical information from earthquakes, analyzes the needs of government at all levels and emergency teams after a devastating earthquake, and studies the extraction of emergency core information and the rapid release of emergency information during the fast response period. Reference [6] has developed a geographic information extraction system based on spatial relationship query. The system uses a geographic information extraction model and can efficiently handle the query of emergency keywords. The above research focuses on fragments of emergency geographic information which include extraction, matching, and querying [7]. Some experts also study the query and index of a class of emergency disaster events [8]-[11].

However, we note that the emergency information in the network is mainly in the form of events as a unit. Events' frame technology is a formal expression of the integrity of the event. Thus, this paper puts forward a technique based on the events' frame of the multisource of emergency geographic information extraction. The events' frame technology is used to construct the emergency event extraction model, the complete method of toponymic recognition and spatialization. Finally, the accuracy evaluation model was used to evaluate the results.

2. System Framework

First, the emergency information extraction model is used to extract the original emergency event information. And, the extracted results are subjected to structured processing. Second, the address recognition method is used to identify the geographical segment. And, the complete address gets the spatial information [12]. Finally, the results are evaluated using the accuracy assessment. The user can query by entering a query statement [13]. Based on the events' frame of the multisource of emergency geographic information extraction shown in Figure 1. There are three levels from bottom to top, including the data layer, logical processing layer, and front display layer.

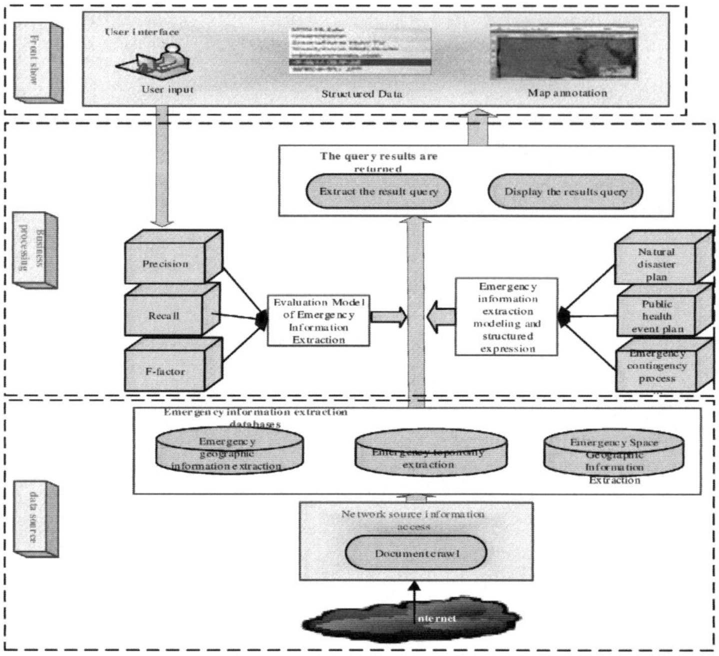

Figure 1. Crowdsourcing information extraction based on events' frame.

3. Key Technology Implementation

3.1. Emergencies Information Extraction Model

The emergency event extraction model is the core module of the events' frame. Its main function is to analyze the extraction of various types of emergency events. Then, it extracts comprehensive information about the incident. Finally, it will extract the results of the structured expression and storage.

The specific process of the extraction method based on emergency events is as follows (refer to Figure 2).

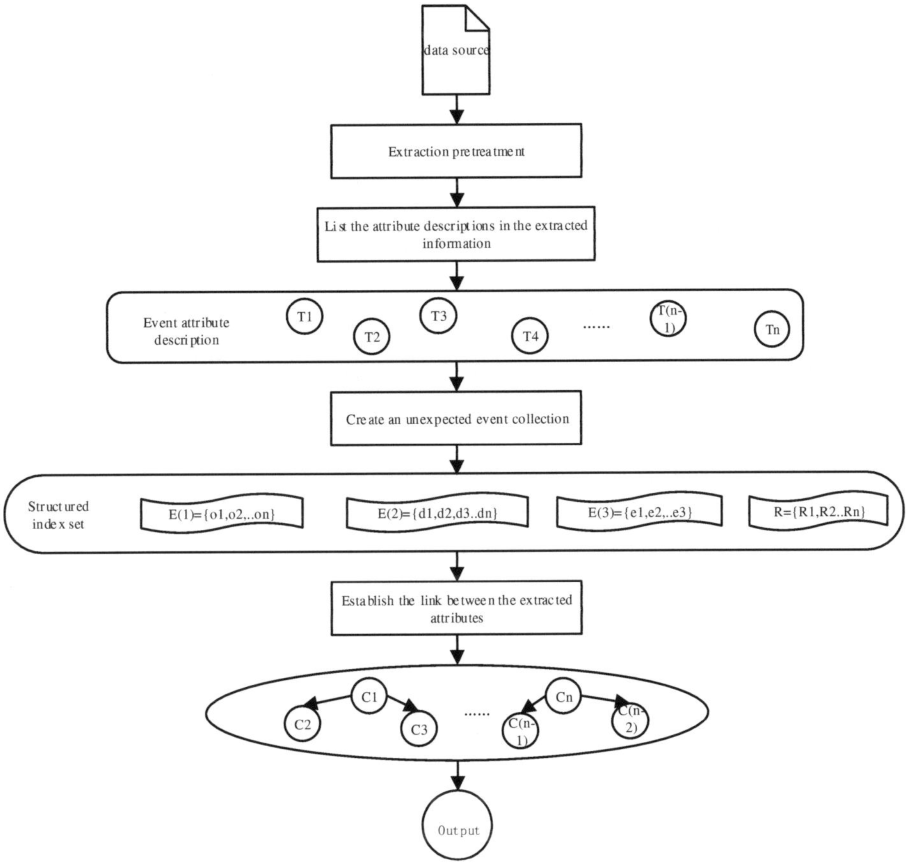

Figure 2. Implementation of emergency information extraction.

3.1.1. Preprocess the contents of the emergency information extraction

The structured content is preserved which include title, type, level, longitude, latitude, time, source, detail, storage, etc.

3.1.2. List the attribute description of the extracted information

According to the emergency information extraction results, the establishment of emergency events and attributes of the relationship between the documents.

3.1.3. Establish a set of emergency events

According to the emergency response to the comprehensive extraction of structured organization and management, the information of the emergency event attribute which the discretization of the extracted result builds a structured set of texts indexes. Through the above three steps to form a contingency geographic information structure of the system.

3.1.4. Establish the link between the extracted attributes

The relationship between the extracted attributes is established according to the relationship in the index sets.

IOP Conf. Series: Journal of Physics: Conf. Series **933** (2017) 012009 doi:10.1088/1742-6596/933/1/012009

We conduct a case study of a series of emergency incident extraction processes [14] which include natural disaster, accident disaster, public health, safe society, etc. The following figure shows the extraction model for emergency events (refer to Figure 3).

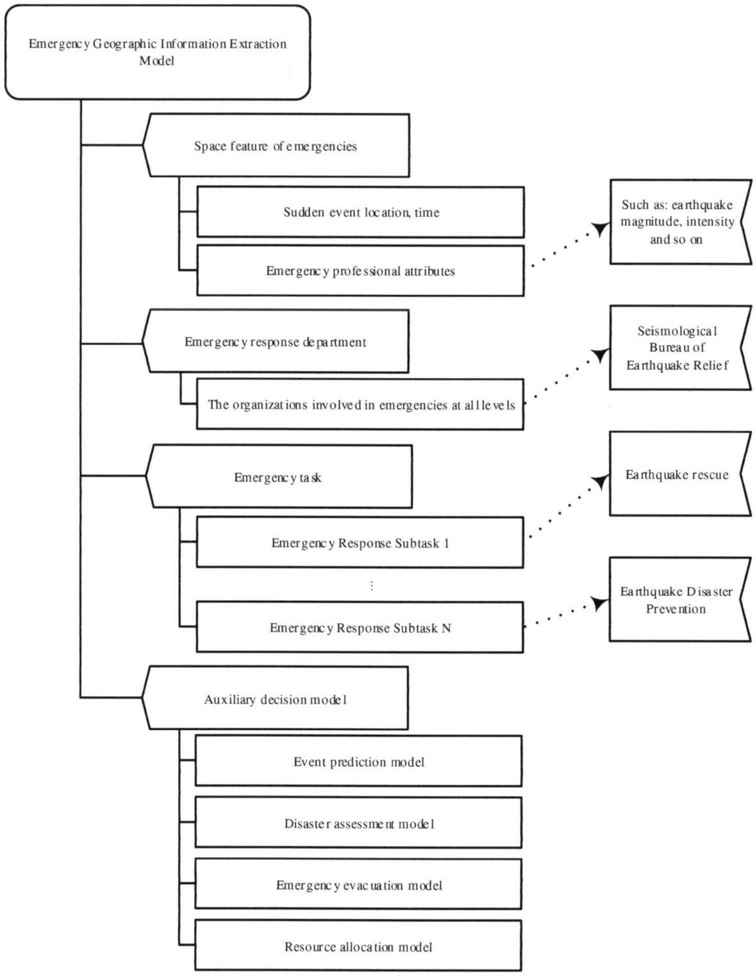

Figure 3 Emergency information extraction model.

The model consists of four elements:

3.1.5. Temporal and Spatial Characteristics of Emergencies

The description of the basic information about the incident, which includes the basic information and professional attributes of emergencies, such as earthquake magnitude and intensity.

3.1.6. Emergency Department

These are the subjects of the various tasks at all levels of emergency response process. This is mainly at all levels of government agencies.

3.1.7. Emergency Task

The sum of the subtasks taken during the emergency response process. Each task is a time and space within a range to achieve a certain purpose of a series of actions integrated.

3.1.8. Auxiliary Decision Model

These models provide supplementary information for task execution which includes disaster prediction model, disaster assessment, emergency evacuation and resource allocation.

3.2. Full-toponomy Recognition Technology Based on a Hierarchical Model

Toponomy recognition based on hierarchical model is the key technology of emergency information extraction model. The hierarchical model is constructed according to China address standard. China's address is decreasing and the scope is from large to small, including the state, province, city, county, district, etc. According to the results of the emergency event to those containing the toponomic fragments of the data, the main function of the technology is to identify the complete address based on the emergency information. The address recognition process based on the hierarchical model is shown in the following figure (refer to Figure 4). This section will cover the main technical aspects of address recognition, including the hierarchical model and the N-shortest path algorithm.

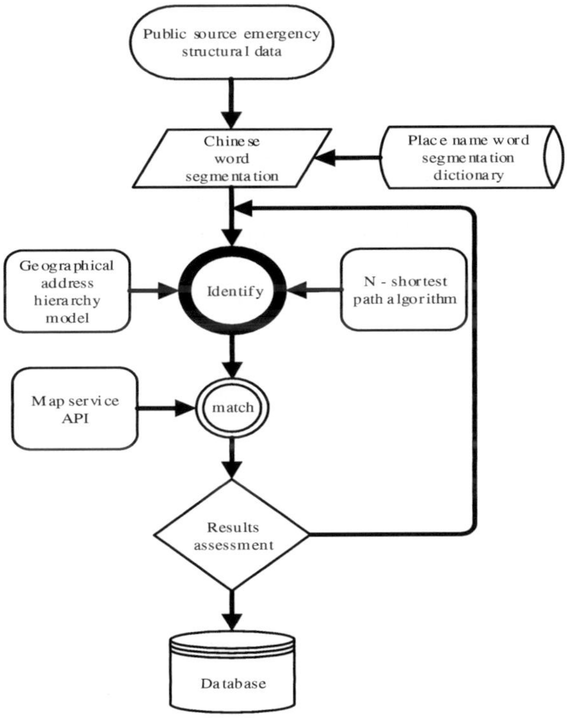

Fig.4 The technical process of place name recognition.

On the basis of referring to the rules of toponomy in China, this paper puts forward the hierarchical toponomy model. The hierarchical model of the identification process is divided into three cases. The hierarchical model only identifies names in national standards. When the toponomy is outside the national standard, the operation of this article is the log output of the name. When the hierarchical model recognition event is partially recognized and another part is not recognized, this article stores the identified content and will be unrecognized content log output. Finally, this article analyzes all the logs. If the toponomy in the log is the missing content of the database, then this article will add the missing content to the corresponding part, otherwise, do not do statistics (refer to Figure 5).
The hierarchy model is defined as follows:

1) <standard toponomy>:=<region><address>

2) <Region>::=[country][province][city][county]

3) <Address>::=road name|community name|place name|building name[floor number|room number]

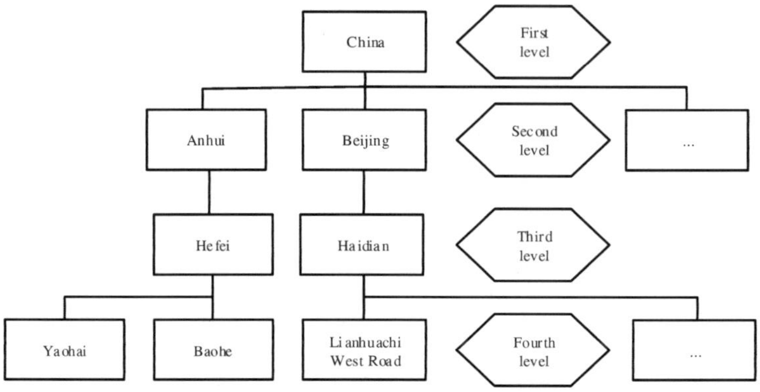

Fig.5 Geographical address hierarchy model.

The hierarchical model splits the names into strings with four elements and does not strictly define the semantics of each layer. In practice, without affecting the efficiency of extraction and identification, the first four are regional geographical names. This is the national, provincial, city, county, district. And the same level of provincial and municipalities, autonomous regions, and special administrative regions. A large number of cases studies found that the four-tier model to meet the demand. In general, there is a spatial inclusion relationship between layers and layers and a space distance on the adjoining relationship. The hierarchical model needs to take full account of the spatial relationship between the toponomys [15]. Spatial relationship refers to the location relationship between the names of elements, including, adjacent, adjoining, azimuth and distance. The model takes the strategy of spatial relation matching. The experimental results show that the spatial constraint data can improve the accuracy of toponymic recognition. The abstraction of the hierarchical model ensures its versatility and strong ability to express the names.

Separate hierarchical models can only connect simple names and cannot achieve full address acquisition. This article uses the N-shortest path algorithm to complete the toponomic fragment real connection [16], [17]. For details, refer to the Dijkstra algorithm [18].

3.3. Evaluation Model of Emergency Information Extraction

On the basis of reference to the evaluation criteria of general information extraction, this paper presents an emergency information extraction and evaluation model. The model is another module of the events' frame. The function of the model is to evaluate the results of emergency information extraction by qualitative and quantitative methods. This paper constructs three aspects of the indicators, namely comprehensive accuracy of emergency geographic information (CAEGI) [19], toponomy recognition accuracy (TRA) [20], [21] and toponomy spatial accuracy (TSA) [22], [23]. The detailed quantitative assessment process is shown below (refer to Figure 6).

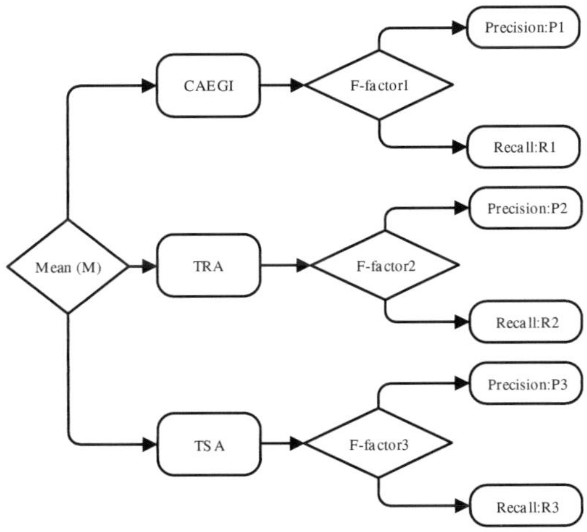

Fig.6 Evaluation model of emergency information extraction.

(1) Precision (P): The ratio of emergency information extraction accuracy to test sample [24].

$$P = \frac{Extract\ results\ (E)}{Samples\ (S)} \times 100\% \tag{1}$$

(2) Recall (R): The correct extraction result and the extraction accuracy ratio [25].

$$R = \frac{Correctnumber(C)}{E} \times 100\% \tag{2}$$

(3) F-factor (F): Precision and recall rate of the weighted average. F is the overall average of the system's extraction performance [26].

The formula for calculating the harmonic mean is:

$$F_\beta = \frac{(\beta^2+1)(P+R)}{\beta^2 \times P + R} \tag{3}$$

The β is the weighting coefficient. The three phases of emergency information extraction are critical. This article treats the importance of precision and recall rates equally. So, the weight coefficient of F is set to 1 in this paper. P is the precision rate; R is the recall rate.

(4) Mean synthetic average (M): M is the expectation of F [27].

$$M(F) = \sum_{i=1}^{n} f_i p_i \tag{4}$$

f is the F-factor calculated by (3). p is the weight of the emergency information extraction result. In the order of extraction, the weights are 1/3, i=1,2,3.

As shown in Figure 6, we use (1) to get the P values for CAEGI, TRA, and TSA. We employ (2) to acquire the R values for CAEGI, TRA, and TSA. We apply (3) to obtain the F values for CAEGI, TRA, and TSA. Finally, we exploit (4) to achieve M results for CAEGI, TRA, and TSA.

4. Results and Discussion

Driven by the emergency events' frame, the emergency information extraction results are shown in the figure (refer to Figure 7). In this paper, 114284 nonempty original web pages are used as test samples.

Test samples from China National Emergency Broadcasting Network. As shown in the following table (see Table 1), the data sources of the site are highly reliable, authentic and real-time update of emergency data.

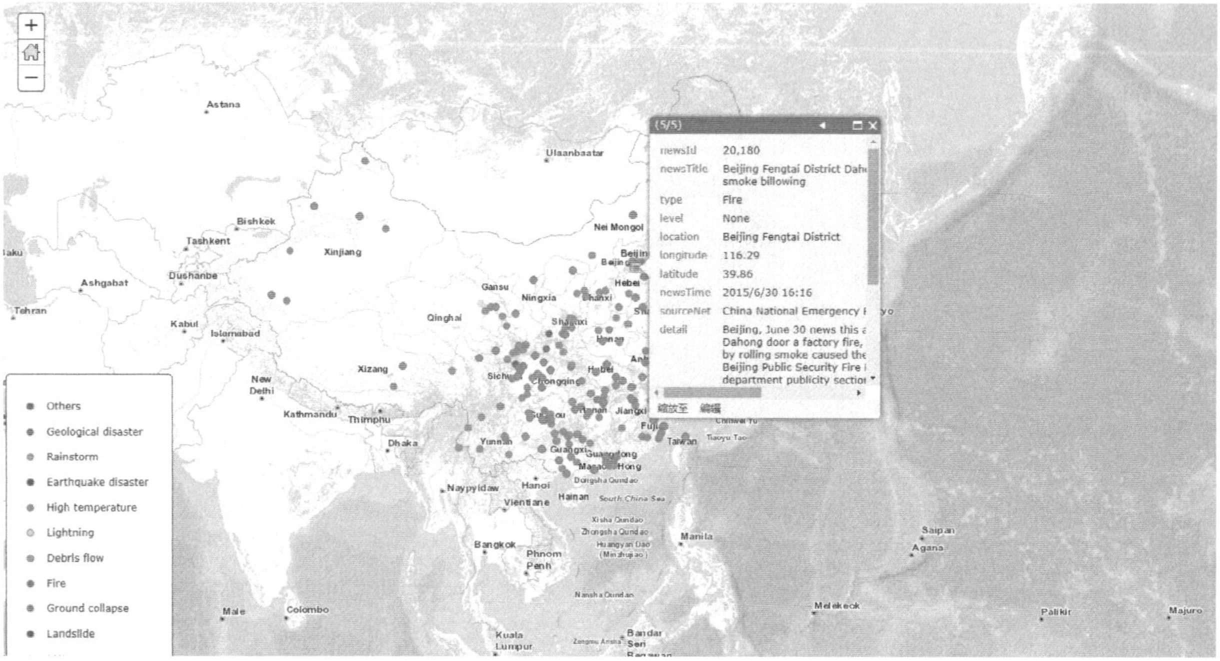

Figure 7. Emergency information extraction results.

Table 1. Case Study on Evaluation Model of Emergency Information Extraction Result

Type	S	E	C	P	R	F
CAEGI	114284	113662	113255	99.10%	99.64%	99.37%
TRA	113662	92846	92808	81.65%	99.96%	89.88%
TSA	92846	90589	90566	97.54%	99.97%	98.74%
M	-	-	-	92.76%	99.86%	95.99%

Notes: Please refer to heading 3.3 on the abbreviations for the table.

The results of the analysis show that the main error of CAEGI is derived from the noise of the test data. There are website science articles and pictures on the site, which pages do not contain incident information. TRA's error is mainly that some pages do not contain toponomic information. At the same time, this test sample is a Chinese page. Chinese segmentation dictionary does not completely cover foreign names information [28]. This is the biggest source of error for TRA. TSA is the main error of TRA caused. TSA is using a domestic map service. When the query to foreign TSA, the error appeared. At the same time, some of the contents of TSA cannot be found [29]. The results of the three types of measurements show that the comprehensive extraction accuracy based on the emergency information events' frame is 95.99%. The emergency geographic information extraction system is reliable and conforms to the actual engineering needs.

5. Conclusion and Future work

There are many information extraction manners exist today, and it is an unattractive operation for us to select the appropriate way. The main purpose of this study is to propose a suitable procedure for emergency geographic information extraction. Event frame technology which including data, models and show layer is for this target. The datasets come from crowdsourcing. The function of the EIEM in the framework is the structural extraction of the emergency information, perfectly matching the address and accurate access to the spatial information of location. The effect of the AEMEI in the frame is to evaluate the extraction results of the EIEM. The final result which meeting the AEMEI is displayed. This framework provides a one-stop operation for the extraction of emergency information. There are still shortcomings in this paper. First, in the EIEM, different data sources need to compile distrinct rules to extract the emergency information. This makes the regulations too time consuming. In the future, we want to use machine learning to automatically draw information. Second, in the TSA, we get the convenience of querying spatial coordinate. For those which cannot query the spatial location information, this article does not propose a fitting method. Third, this article returns the results based on the user's query. We should use the recommended system to intelligently decision the results. For the future work, this is still the direction of further research.

6. References

[1] Yang, J. (2017). *Environmental Emergency Response Plan. Environmental Management in Mega Construction Projects*. Springer Singapore.

[2] Etzioni, O., Fader, A., Christensen, J., Soderland, S., & Mausam, M. (2011). Open Information Extraction: The Second Generation. IJCAI 2011, *Proceedings of the, International Joint Conference on Artificial Intelligence, Barcelona, Catalonia, Spain, July* (Vol.14, pp.3-10). DBLP.

[3] Yang, P., Wang, W., & Dong, C. (2009). Application of Emergency Case Ontology Model in Earthquake. *International Conference on Management and Service Science* (pp.1-5). IEEE.

[4] Jiang, D. L. (2010). Research on extraction of emergency event information based on rules matching. *Computer Engineering & Design*, 31(14), 3294-3297.

[5] Min, L. I., Zhao-Long, L. I., Huo, J., & Yong-Qiang, L. I. (2015). Research on technology of extraction and rapid release of key information about earthquake emergency. *Journal of Seismological Research*.

[6] Kalashnikov, D. V., Ma, Y., Mehrotra, S., Hariharan, R., & Butts, C. (2006). Modeling and querying uncertain spatial information for situational awareness applications. *ACM International Symposium on Geographic Information Systems, Acm-Gis 2006, November 10-11, 2006, Arlington, Virginia, Usa, Proceedings* (pp.131-138). DBLP.

[7] Bergen, J. R., Anandan, P., Hanna, K. J., & Hingorani, R. (1992). Hierarchical Model-Based Motion Estimation. *European Conference on Computer Vision* (Vol.11, pp.237-252). Springer-Verlag.

[8] Grishman, R. (2015). Information extraction. *Foundations & Trends in Databases*, 30(5), 8-15.

[9] Cowie, J., & Lehnert, W. (1996). *Information extraction. Information extraction* :. Springer.

[10] Cowie, J., & Lehnert, W. (1996). Information extraction. *Foundations & Trends in Databases*, 1(3), 261-377.

[11] Wang, J., Yan, W., & Ngueyn, M. (2017). *A Framework of Event-Driven Traffic Ticketing System*. IGI Global.

[12] Kurra, S. (2014). Solar power plant design and location selection for tamuk using gis and nlp. *Dissertations & Theses - Gradworks*.

[13] Downey, D., Etzioni, O., Soderland, S., & Weld, D. S. (2004). Learning text patterns for web information extraction and assessment. *National Conference on Artificial Intelligence* (pp.50--55).

[14] Seymore, K., Mccallum, A., & Rosenfeld, R. (1999). Learning hidden markov model structure for information extraction. *Proc Aaai'99 Workshop Machine Learning for Information Extraction*, 37--42.

[15] Bian, F., & Wan, Y. (2009). A novel spatial co-location pattern mining algorithm based on k-nearest feature relationship. *Geomatics & Information Science of Wuhan University*, 34(3), 331-334.

[16] Johnson, D. B. (1973). A note on dijkstra's shortest path algorithm. *Journal of the Acm*, 20(3), 385-388.

[17] Zheng, F., Simpson, A. R., & Zecchin, A. C. (2011). A combined nlp-differential evolution algorithm approach for the optimization of looped water distribution systems. *Water Resources Research*, 47(8), 2924-2930.

[18] Mehlhorn, Schäfer, & Tamaki. (2003). A heuristic for dijkstra's algorithm with many targets and its use in weighted matching algorithms. *Algorithmica*, 36(1), 75-88.

[19] Yandell, M. D., & Majoros, W. H. (2002). Genomics and natural language processing. *Nature Reviews Genetics*, 3(8), 601.

[20] Jagannathan, V., Mullett, C. J., Arbogast, J. G., Halbritter, K. A., Yellapragada, D., & Regulapati, S., et al. (2009). Assessment of commercial nlp engines for medication information extraction from dictated clinical notes. *International Journal of Medical Informatics*, 78(4), 284-291.
[21] Cogley, J., Stokes, N., & Carthy, J. (2013). Exploring the effectiveness of medical entity recognition for clinical information retrieval. *International Workshop on Data and Text Mining in Biomedical Informatics* (Vol.5, pp.3-4). ACM.
[22] Okoye, I., Mahmud, J., Lau, T., & Cerruti, J. (2011). Find this for me: mobile information retrieval on the open web. *International Conference on Intelligent User Interfaces* (pp.3-12). ACM.
[23] Willis, A. (2015). Using NLP to Support Scalable Assessment of Short Free Text Responses. *Tenth Workshop on Innovative Use of Nlp for Building Educational Applications* (pp.243-253).
[24] Dimitriadis, S., & Goumopoulos, C. (2008). Applying Machine Learning to Extract New Knowledge in Precision Agriculture Applications. *Panhellenic Conference on Informatics* (pp.100-104). IEEE Computer Society.
[25] Jijkoun, V., Rijke, M. D., & Mur, J. (2004). Information extraction for question answering: improving recall through syntactic patterns. *International Conference on Computational Linguistics* (pp.1284). Association for Computational Linguistics.
[26] Makhoul, J., Kubala, F., Schwartz, R., & Weischedel, R. (1999). Performance measures for information extraction. *Proceedings of Darpa Broadcast News Workshop*, 249--252.
[27] Pfeiffer, P. E. (1990). Mathematical Expectation. *Probability for Applications*. Springer New York.
[28] Florczyk, A. J., Nogueras-Iso, J., Zarazaga-Soria, F. J., & Béjar, R. (2012). Identifying orthoimages in web map services. *Computers & Geosciences*, 47(4), 130-142.
[29] Abdulkarim, B., Kamberov, R., & Hay, G. (2014). Supporting urban energy efficiency with volunteered roof information and the google maps api. *Remote Sensing*, 6(10), 9691-9711.

Acknowledgment

This work was supported by the Incident Information Fusion and Visualization Based on Public Source Data (Grant No. 7771615).

Bo Yang was born in Anhui, China, in 1992. Currently, he is pursuing his M.S. degree at the School of Mapping and Geographical Science, Liaoning Technical University, Liaoning, China, from September 2015 to June 2018. His research interests include data mining, geographical information science.

Jizhou Wang was born in Hebei, China, in 1976. He obtained his Ph.D. in cartography and GIS at Wuhan University in 2005. He leads the local GIS research group which focuses on research in the field of geographical information systems.

Weijun Ma was born in Hebei, China, in 1976. He obtained his M.S. in cartography and GIS at Chinese Academy of Surveying & Mapping in 2004. His current research interests include population GIS, especially spatialization of population, constructing population information grid and census, and data mining, emergency surveying and mapping.

Xi Mao was born in Zhejiang, China, in 1983. He obtained his B.S. Ph.D. in cartography and GIS at Wuhan University, Peking University in 2005, 2010, respectively. His current research interests include urban geospatial modeling, geospatial information search engine, geospatial information services.

IOP Conf. Series: Journal of Physics: Conf. Series **933** (2017) 012010 doi:10.1088/1742-6596/933/1/012010

AI based HealthCare Platform for Real Time, Predictive and Prescriptive Analytics using Reactive Programming

Jagreet Kaur[1], Dr. Kulwinder Singh Mann[2]

[1]Data Scientist, XenonStack, IT Park, Chandigarh, India.
[2]Professor and Head of the IT Department, GNDEC, Ludhiana, India

Abstract. AI in Healthcare needed to bring real, actionable insights and Individualized insights in real time for patients and Doctors to support treatment decisions., We need a Patient Centred Platform for integrating EHR Data, Patient Data, Prescriptions, Monitoring, Clinical research and Data. This paper proposes a generic architecture for enabling AI based healthcare analytics Platform by using open sources Technologies Apache beam, Apache Flink Apache Spark, Apache NiFi, Kafka, Tachyon, Gluster FS, NoSQL- Elasticsearch, Cassandra. This paper will show the importance of applying AI based predictive and prescriptive analytics techniques in Health sector. The system will be able to extract useful knowledge that helps in decision making and medical monitoring in real-time through an intelligent process analysis and big data processing.

1. Introduction

As with the change in the time and advancements in the technology, there is a need to make a systematic change to health systems to improve the quality, efficiency, and effectiveness of patient care. Chronic diseases like heart disease, stroke, cancer, and diabetes are considered as the most common, expensive, and preventable health problems but due to the poor health care systems, patients can't able to take good care of the problems.

The strategic aim of value based Health Care is to ensure that everyone can use the health services they are needed for their good health and well-being. The focus on value-based care corresponds to an increased concentration on patient-centric care. By focussing on technologies and healthcare processes on patient results, a doctor, hospitals, and health insurance need to work with each other to personalize care that is effective, transparent in its delivery and billing, and measured based on satisfaction of the patient.

Suppose a patient is suffering from ache or pain and need to visit a physician. After listening to patient symptoms, Physician inputs them into the computer, which helps to show the latest research that is needed to know about how to diagnose and treat the problem. Patient has an MRI or an x-ray and a computer helps to detect a problem that could be too small for a human to see. Finally, a computer monitors at patient medical records and family history and compares both with the most recent research to suggest a treatment for a patient that is specifically personalized to his needs.

Nowadays, Patients need an affordable and high quality of healthcare. According to the Health Data, most of the information is not in structured, relational format. About 80% information is in an unstructured format. Due to the limited structured data about the patient's health conditions is available and patients have only very limited opportunities to actively involved in the process. It is very difficult to use this massive unstructured amount of data from different sources to take the right decision for the right patient at the right time by the doctors. This will slow down the personalized

care to the patient. So, it is necessary to develop a new strategy or system to care for patients which involves the health of patients while decreasing the cost of care[1].
These services range from clinical care to the public services for individual patients that are helpful for the health of whole populations. There is a need to improve the healthcare quality and coordination so that outcomes are consistent with current professional knowledge. The cost of treatment for the problems should be reduced so that each and every patient get their personalized treatment at lower cost.

Adoption of Electronic Health Records and systematic collection of data by health care providers were predicted to improve the efficiency and quality of patient care.
Machine learning is improving diagnostics, predicting outcomes, and just beginning to scratch the surface of personalized care. Human surgeons will not be replaced by machines in the future, but AI can positively help surgeons to make better clinical decisions or can even replace human decision in certain areas of healthcare (eg, radiology). The growing availability of healthcare data and speedy development of big data analytic methods have made possible the current successful applications of AI in healthcare [2].
Data analytics has become increasingly important in almost every region of the economy. Health care involves a diverse set of public and private data collection systems having different sources of Data. The volume of data has been increasing exponentially in the past era as health care contributors turned to EHRs, digitized laboratory slides, and high-resolution radiology videos and images. There are petabytes of data stored in databases of health insurance companies and the trillions of data points streaming from sensors like activity trackers and various other continuous monitoring devices.
As shown in fig1. Text or natural language data resides in many fields including several fields such as output from medical devices, physician notes, nursing notes, surgical notes, radiology notes, pathology reports, admission notes, clinical data, Genomic data, Behaviour data etc. These fields may have valuable information about the patient including diagnosis, history, family history, complaints, statistics, and opinions, demographics, medical history, allergies, laboratory test results etc.

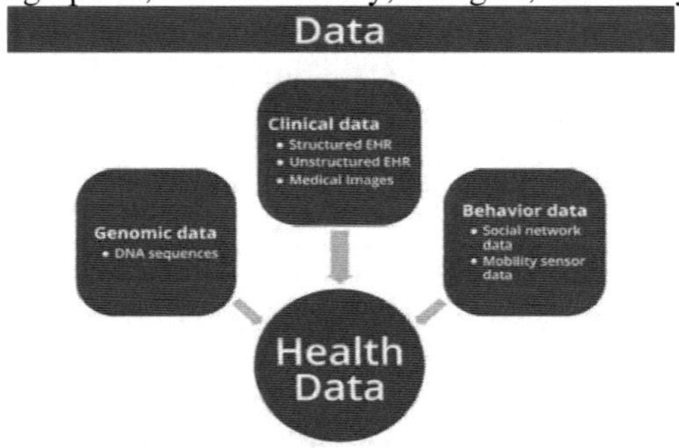

Figure1. Health care data sources

To transform the current healthcare system into a preventive, active and value-based system, the interoperability, exchange, and sharing of health data are needed.
One of the major challenges is data integration i.e. to integrate the data that is obtained for each patient into one system, as that will allow for fast data analysis, and give physicians all the information they need to perfectly treat their patients [3]. However, most of the data is encrypted and access is restricted due to the patient privacy and many medical devices are not interoperable[4]. So, once a single database can be established, ML can solve the major problems. Machine learning provides a way to find reason and patterns about data, which enables physicians to move to personalized care known as precision medicine. There are many ways as how machine learning algorithms can be used in healthcare, but all of them depend on having sufficient amount of data and permission to use it. Previously, alerts and advise for a medical treatment have been developed based on physicians studies, and data into their software. However, that can restrict the accuracy of that data because it might come from different populations. On the other hand, Machine learning nan be refined using data that is available in that particular environment i.e. anonymous patient information from a hospital and the

IOP Conf. Series: Journal of Physics: Conf. Series **933** (2017) 012010 doi:10.1088/1742-6596/933/1/012010

area it serves. Physicians can use machine learning to predict hospital readmission for chronically ill patients. The patients that are at risk of being readmitted are identified that makes possible for providers to offer improved post-discharge support. By lowering the rate of readmission, it can improves the lives of those most at risk.
The rest of this paper is organized as follows: Section II gives an overview of Interdependency of Cloud Computing with Big Data Technologies. Section III describes about the Bigdata in terms of Healthcare. Section IV the related work. Section V describes the platform design in which steps needed to do analytics will be discussed. Section VI describes the architectural design. Section VII gives an overview about artificial intelligence that helps the system to understand from its experience and reactive machine learning.

1.1. Need of Health care system

- Reduce readmissions. Health care systems can reduce readmissions in an efficient and patient-centred manner. Clinicians can receive daily guidance as to which patients are readmitted and how they reduce that risk.
- Prevent hospital acquired infections (HAIs). Health systems can reduce HAIs, such as central-line associated bloodstream infections (CLABSIs)—40 percent of CLABSI patients die—by predicting which patients with a central line will develop a CLABSI. Clinicians can monitor high-risk patients and intervene to reduce that risk by focusing on patient-specific risk factors.
- Reduce hospital Length-of-Stay (LOS). Health systems can reduce LOS and improve other outcomes like patient satisfaction by identifying patients that are likely to have an increased LOS and then ensure that best practices are followed.
- Predict chronic disease. Machine learning can help hospital systems identify patients with undiagnosed or misdiagnosed chronic disease, predict the likelihood that patients will develop chronic disease, and present patient-specific prevention interventions.
- Predict no-shows. Health systems can create accurate predictive models to assess, with each scheduled appointment, the risk of a no-show, ultimately improving patient care and the efficient use of resources [5].

2. Interdependency of Cloud Computing with Big Data Technologies

Healthcare is moving towards the digital platform, more patient-oriented and data-driven. Availability of data, irrespective of the location of the patient and the clinician, has become the key to both patient satisfaction and improved clinical outcomes. Cloud technologies can significantly facilitate this trend.
Healthcare functionality can be improved by cloud-based healthcare IT systems that provide the potential for broad interoperability and integration. Healthcare cloud services are Internet-based and use standard protocols, so connecting them to other systems and applications is very simple. The ability to share information easily and securely is a critical capability, and cloud services are good enablers for this[6].
Healthcare data needs to be shared across various settings and geographies which further burden the healthcare provider and the patient causing a significant delay in treatment and loss of time. Cloud services can permit remote access to applications and data[7]. The Internet connectivity enables access at anytime from anywhere. Cloud caters to all these requirements thus providing the healthcare organizations an incredible opportunity to improve services to their customers, the patients, to share information more easily than ever before, and improve operational efficiency at the same time [8].
Cloud computing has special features for physicians and patients that help to reduce the burden of heavy investments and to utilize outsourced resources, software, hardware, automated resource management, parallel computing, and utility computing[9]. The objectives of such usage include improvement in patient care, continuous feeds of physiological data are analyzed, financial processes are optimized and resource utilization.

2.1. Big Challenges in Health care system

- No unified format: As there is no standard method adopted in hospitals, labs to generate reports etc., they generate reports in their own formats so it is a challenge to integrate different formats and process the data.
- Large volume of data: If we take the case of image data like X-rays etc., a large size of data is generated for a single patient and there is a need to efficiently handle such data and extract useful information.
- Real time data handling: Need to capture the patient's behavioral data through several sensors, social interactions and communications.
- Integration and sharing of data: Efficient data integration and sharing rely on standardized coding schemes and terminologies. Currently, standardized coding systems are used for high-level information coding only. A lot of information is not coded properly. In order to data integration and sharing, coding schemes, as well as data models need to be improved and standardized[10].

2.2. Real-Time Analytics & Predictive Analytics: - Knowledge Discovery

To take the Right decision for the Right patient at the right time by the doctors, Doctors/Physicians need knowledgeable Data, Real-Time Analytics, and Predictive Modelling to improve the health care [11]. Real-time healthcare analytics can help to improve the quality of care, cost, and meet various requirements by streamlining and automating the process of collecting and measuring vast amounts of healthcare data.

Too often today, there is a need to handle multi-sourced and multi-typed of data, including unstructured and semi-structured data and relational databases like SQL aren't capable of handle this. Also, need to analyze terabytes or petabytes of data having response times of sub seconds and it is difficult for a single server to handle this. Last, batch processes are needed for some jobs. But many times, we need to analyze rapidly changing, multi-structured data in real time. It is very difficult to do Extract, Transform and Load(ETL) processes to cleanse data at that time. Real-time analytics tools handle these issue by bringing dissimilar information from many sources into the one place. Real-time healthcare analytics can help to improve the quality of care, reduce costs, and help to meet regulatory requirements by automating the process of collecting and measuring huge amounts of healthcare data. It can specify vision into historical, current, and predictive disease and treatment of unaddressed conditions that need the physician's attention so that problems can be addressed before the patient is readmitted and the hospital is penalized.

Knowledge discovery and Data mining are two distinct terms. Knowledge Discovery is a kind of process used for extracting the useful information from the database and Data Mining is one of the steps involved in this process. Data Mining is the algorithm used for extracting the patterns from the data.

Knowledge Discovery involves various steps such as Data Cleaning, Data Integration, Data Selection, Data Transformation, Data Mining, Pattern Evaluation, Knowledge Presentation. Knowledge Discovery is a process that has total focus on deriving the useful information from the database, interpretation of storage mechanism of data, implementation of optimum algorithms and visualization of results. This process gives more importance on finding the understandable patterns of data that further used for grasping useful information.

Knowledgeable data is basically the knowledge extracted from information. It is essential for healthcare organizations to effectively manage both internal knowledge and externally generated knowledge to provide the best possible health care facilities to the patients. It increases decision-making capabilities of Physicians [12].

Predictive modeling is a technique to predict in advance a disease of a patient so that diagnose of the problem can be done in advance. Predictive Modelling benefit to large Healthcare providers to discover the treatment for personalized care managing large patient populations. Predictive analyses can be done by inputting the EHR data into a data model that was processed by various analytic techniques and analyze the results. The rich sources of data along with analytics capabilities have potential for an increased understanding mechanisms of disease and better health care.

3. BigData in terms of Healthcare sector

Big Data refers to the vast quantities of data generated from the Internet and digitization of all sorts of information, including health records[13]. New big data technologies hold commitment for centralizing and analyzing these digital data in order to discover trends and make predictions. Big data is commonly characterized using a number of V's. The first three are volume, velocity, and variety. Volume refers to the vast amounts of data that is generated every second, minutes, hour, and day in our digitized world. Variety refers to the ever increasing different forms of data that can come in such as text, images, voice. Velocity refers to the speed at which data is being generated and the pace at which data moves from one point to the next. Volume, variety, and velocity are the three main dimensions that characterize big data and describe its challenges[14]. More Vs has been introduced to the big data community as we discover new challenges and ways to define big data are veracity, visualization, viability, vulnerability, volatility, and value. Veracity refers to whether the data is trustable. Visualization refers to the way by which data is presented to the user. Viablilty refers to the relevance of the data. Vulnerability refers to the security of the data. Volatility refers to the change of data with the time Last, Value refers to check the data for the meaningful results.

3.1. Big Data trends in Healthcare Sector

Evidence-based medicine should be given, which involves a use of all clinical data and factoring that into clinical and advanced analytics. To capture and bring all information of patient together is difficult but it gives a complete view for understanding into care coordination and population health management, and patient engagement and outreach.

The improvement of the quality of care can be focussed if the various extents of health data are combined in the automated health data analysis. The data dimensions include (a) the clinical data that describes the health status and history of a patient, (b) the clinical and administrative clinical process data, (c) the information about diseases as well as related population data, and (d) the awareness about changes. If the data analysis is limited to only one data feature, for example, the administrative and financial data, it will not be possible to identify new standards for personalized treatments. So, the integration of the various heterogeneous datasets and exchange of health data is an important requirement of big health data applications.

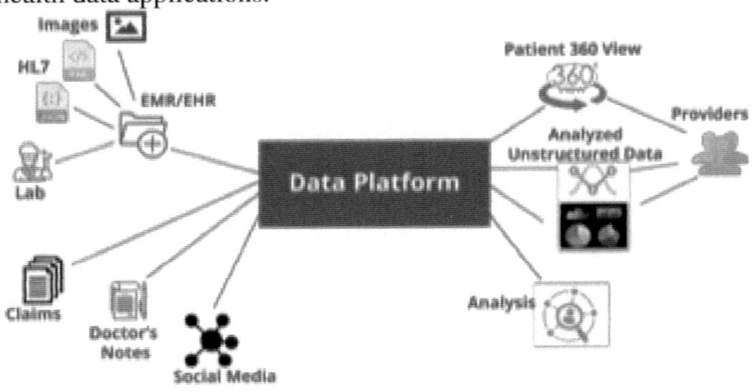

Figure 2 Data Platform

3.2. Key Elements of Healthcare Sector

Integrating Data: To understand every possible feature about customers, citizens, and clients to apply advanced analysis and computation to modify existing strategies, there is a need to integrate a large set of data and business analytics. In a similar manner, To identify new patterns and to improve healthcare by finding the right treatment for an individual's, there is a need to link heterogeneous data sets securely.

IOP Conf. Series: Journal of Physics: Conf. Series **933** (2017) 012010 doi:10.1088/1742-6596/933/1/012010

Generating New Knowledge: The primitive uses of Big Data were to generate new visions through predictive analytics. In addition to the clinical and administrative data, integrate patient data and their environment that helps in better predictions and right treatments to the right patients.
The predictions may also help in identifying the areas that need improvement such as treatments, early identification of worse health states, readmissions, etc.
Transforming Knowledge into Practice: Although the standardization of data collection and fresh analytical approaches to the big data revolution in healthcare are important, it's their practical application that will get it across the line.
The understandings that are obtained from big data have the potential to reform multiple extents of healthcare like the outcomes from comparisons, the effectiveness of various treatments, and the predictive models that can be used for diagnosing, treating, and delivering care.

4. Related Works

Now days, Healthcare analytics is a new trend in the field of analytics. Due to the advancements in technologies in Health care sector, there is a major breakthrough in data collection. To improve the processing of predictable healthcare system, we have proposed a series of Big Data Health Care System. There are many techniques proposed in order to proficiently process large volume of medical record. Akshay Raul, Atharva Patil, Prem Raheja, Rupali Sawant proposed "Knowledge Discovery, Analysis and Prediction in Healthcare using Data Mining And Analytics"[16]. In this paper, Author proposed a Healthcare system that will be used to create a public awareness about alternative drugs for a specific medicine, availability of that alternative medicine in an area. The Proposed system helps the patients to find an alternate medicine which is prescribed by the doctor. Aditi Bansal and Priyanka Ghare proposed "Healthcare Data Analysis using Dynamic Slot Allocation in Hadoop". In this paper HealthCare System is analysis using Hadoop using Dynamic Hadoop Slot Allocation (DHSA) method[17]. This paper proposed a framework which focus on improving the performance of MapReduce workloads and maintain the system. DHSA will focuses on the maximum utilization of slots by allocating map (or reduce) slots to map and reduce tasks dynamically. Van-Dai Ta, Chuan-Ming Liu, Goodwill Wandile Nkabinde proposed "Big Data Stream Computing in Healthcare Real-Time Analytics"[18]. In this paper, Author proposed a generic architecture for big data healthcare analytic by using open sources, including Hadoop, Apache Storm, Kafka and NoSQL Cassandra. Wullianallur Raghupathi and Viju Raghupathi has proposed "Big data analytics in healthcare: promise and Potential"[19]. In this paper author proposed the potential of big data analytics in healthcare. The paper provides an overview of big data analytics for healthcare practitioners and researchers. Big data analytics in healthcare is growing into a promising field for providing vision from very large data sets and improving results while reducing costs. This paper proposes a generic architecture for big data healthcare analytic by using open sources, including Hadoop, Apache Storm, Kafka and NoSQL Cassandra.
However, all these works either consider the techniques to extract features form specific healthcare data sources or only focus on batch-oriented task to compute which has higher latency computing. In our proposed framework, data sources are supposed constantly coming with high rate, variety of formats, and high volume. Stream computing of real data and use of cloud computing in serving layer will enhance the results of healthcare analytics.

5. Platform Design

This demonstrates the major components of the Healthcare Analytical platform and the types of services that is needed to build a model.
Data Ingest: The first problem that everyone gets into is data ingest. The device data is coming from anywhere and of any type, it will be represented in a standard manner to analyze.
In this part of the application, solution of these problems is discussed:
How to ingest data from medical devices?

How to transform the device data into a format that can be analyzed in a Streams application?
What are the common data schema types when analyzing medical device data?

Data Preparation: Next, there is a need to prepare the data to analyze. Some of the common problems need to remove include Deduplication, Resampling, Normalization, Cleaning, Noise reduction etc[20].

Data Processing: Data processing can be done by using two types like Base Analytics and Aggregated analytics.

Base Analytics: The platform should provide a set of basic analytics of data. These analytics can be used as building blocks for the complex analytics and prediction models. Here are some of the things we can do like Simple vital analytics (calculating rolling average, raising alerts when vitals going beyond normal range), Analytics of ECG, EEG, ICP Waveforms etc.

Aggregated Analytics: This is an area where we combine and aggregate results from the base analytics to form more sophisticated analytic rules. For example, for septic shock detection, the user should be able to describe a rule like this:

- if temperature is > 37 degree Celsius or < 33 degrees Celsius
- and if heart rate is > 90
- and respiratory rate is > 19 or PaCO2 < 31 mm HG
- and WBC > 13,000/mm3, or < 5000/mm3, or > 9% band
- then raise an alert for early septic shock detection.

Patient Data Correlation / EMR Integration: For more complex analytics, we may need to use EMR data or doctor's notes. For example, we may need to retrieve patient's medical history. Or need to retrieve doctor's notes that include some of the doctor's observations[21].

These types of data can be ingested from existing hospital infrastructure using the HL7 / FHIR protocol.

Central Monitoring Dashboard: As part of the platform, there is a need to create a simple dashboard. This will help users to visualize their data and validate their analytic results. The dashboard can be web-based, mobile.

Alert and Notification Framework: When an important event occurs, we need to be able to notify and alert the right people to help the patient. In this part of the framework, we need to deliver notification to the right people based on alert types and patient information, alerts can be delivered via email, text messaging, etc. or should be displayed onto a dashboard.

6. Healthcare Platform Architecture

We require to collect the data, process the data, storage of data, and finally serve the data for analysis, machine learning, and dashboards. In this section, we propose an architecture including the advantages of batch and stream computing to enhance big data computing in healthcare. It can also deal with a large data set by providing cost reduction along low latency and better health care conditions.

The fig 2. shows the detailed architecture of Healthcare Platform.

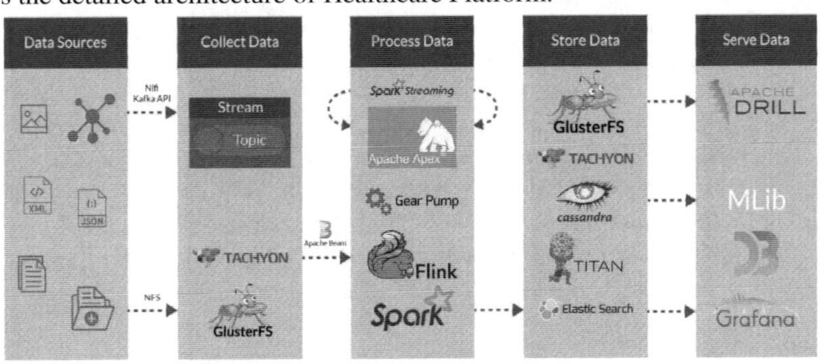

Figure 3. Data Pipelining

Data sources: As there are diverse sources of Health data available now a days like data from medical devices, Social media, EMR, Back- office-systems, Pharmacy, Medical images and HL7 events

Data ingestion: While developing the system, the first step needed is to integrate the data i.e. collect the data from various data sources. Data Ingestion can be from multiple sources, so need a unified platform where we can manage all our data sources. Here the data can be generated from two types of sources, one is real time data and other is batch data. To ingest data from real time data sources, Apache NiFi and Kafka API can be used.

• **Apache NiFi:** Apache NiFi is distributed, scalable, fault-tolerant workflow automation platform. So, we will be having unified view on our Web UI from where we can monitor all our different data sources from where data ingestion is taking place[22].

• **Apache Kafka:** Apache Kafka is a Distributed Messaging System and disk based cache. It can store the data that is collected by Apache NiFi. It provides unified, high-throughput, low latency and works on publish-subscribe model.

To ingest data in batches, NFS (Network File system) can be used. Data from Batch files can be ingested using Apache NiFi and Kafka API also but to process older data that was stored on distributed servers can only be accessed through NFS. NFS protocol provides remote access to shared disks across networks. An NFS-enabled server can share directories and files with clients, allowing users and programs to access files on remote systems as if they were stored locally. An NFS-mounted cluster allows easy data ingestion of data sources such as files, images... from other machines leveraging standard Linux commands, utilities, applications, and scripts.Gluster FS is free and open source software that can be used for distributed storage of images and videos and used for data analysis whereas Tachyon provide in memory storage with fast data processing. It is also used to process the data at memory speed.

Data Processing: After ingestion of data from various sources, next step is to process the data i.e. to clean the data, normalize the data, remove all outliers, data mining, etc. For data processing, Apache spark and apache apex will be used[23].

Apache spark is a next generation distributed parallel processing framework having a rich set of APIs for machine learning, GraphX for graph processing[24]. Spark is much faster than MapReduce for iterative algorithms because Spark tries to keep things in memory whereas MapReduce involves more reading and writing from disk[25] .

Apache Gearpump is a lightweight real-time big data streaming engine. It is a simple pump that consists of two gears. It is difficult to process stream data because of low latency assurance, infinite data, Out-of-order data, • Correctness requirement, Cheap to update user logics but Apache Gearpump solve all these problems as it provides good user interface, control the flow of data, out-of-order processing.

Apache Apex is an engine for processing streaming data. Some other engines that can also fulfil the requirement of stream processing are Apache storm, Apache flink[26]. But Apache Apex has a built-in support for fault-tolerance, scalability and focus on operability.

Apache Flink is an open source platform for distributed stream and batch data processing. Flink has an API that perform operations on streams of data. At the time of submission of a job, theseoperations are turned into a graph by Flink.. It also supports an API that are compatable with storm.

Data Storage: As already discussed, Health care data consists of a variety of data like images, doctor notes, lab reports, insurance data etc. Different types of storage system are also needed like cloud to store the data. To store the data Gluster FS, Tachyon, Cassandra, Titan DB and Elasticsearch can be used. Gluster FS is free and open source software that can be used for distributed storage of images and videos and data analysis. Tachyon is used to process the data at memory speed. Also provide in memory storage with fast data processing. Apache Cassandra is a free and open-source distributed database management system used to handle large amounts of data like sensor data and claim data. It is an open source project developed by Facebook Titan dB is a database used to store graph and link analysis. Elastic search is a search engine used for indexing.

7. AI Based Healthcare System

AI refers to 'Artificial Intelligence' which means making machines capable of performing intelligent tasks like human beings. AI performs automated tasks using intelligence.

Artificial intelligence in healthcare is a new research area that combines representational and computing techniques with the visions of expert physicians to produce tools for improving health care.

It is the study of ideas which allows computers to do the things that make people seem intelligent. The central goals of Artificial Intelligence are to make computers more practical and to understand the principles which make intelligence possible

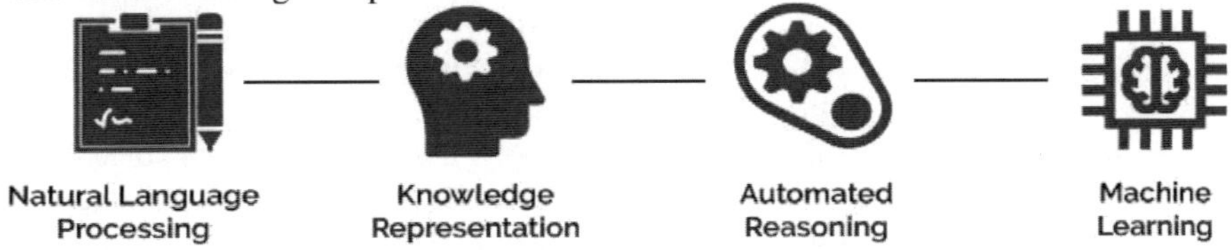

Figure 4. AI enabled Healthcare System

There is need to make system AI enabled so that machine can automatically predict and prescribe the results from its own experience. To make system AI enabled, Natural Language processing, Knowledge representation, Automated reasoning, and Machine Learning should be in the system as shown in fig 3. NLP helps to make computer/machines as intelligent as human beings in understanding language.

7.1. Use Cases of AI based Healthcare System Architecture

One of the most effective use of AI based Healthcare system is to optimize the clinical process. AI based mobile app ask patients about their symptoms and provides an easy to understand information about their Health. The system uses natural Language processing to provide a rich experience and machine learning algorithms to create a map of patient condition and give a personalized experience[27].

AI is helpful in Monitoring patient vitals in real time. In a hospital, nurses do rounds and visit each and every bed after every few hours to manually measure and record patient vital signs. But the patient's condition may drop between the time of planned visits. This means that physicians often respond to problems reactively, in situations where earlier arrival may have improved the patient's wellbeing.

Latest wireless sensors can capture and transmit patient's vitals far more often and these measurements can stream using some various new technologies. The real-time alerts can be generated and Physicians can use these signals and respond more rapidly to unexpected changes[28].

The treatment and prevention of dangerous diseases depends on detecting the symptoms at the right time. In various cases, early diagnosis can result in complete cure. Contrariwise, a late or wrong diagnosis can have damaging results. It is difficult for humans to make reliable decisions.

AI algorithms can rapidly ingest billions of samples in short order and collect useful patterns. Machine learning algorithms can be used to make the knowledge bases used by expert systems and predictive analysis of data. Hence, AI based Healthcare system is helpful for patients as well as doctors.

7.2. Integrating Big Data, Analytics, Artificial Intelligence, and Machine Learning in Healthcare

With enormous amounts of computational power, machines can analyze large sets of data points and apply correlation modelling in a predictive way and in real time. Big Data technology has the potential to influence machine learning capabilities accurately and real-time decision-making capability helps in improving overall operating efficiency and reducing unnecessary costs.

In healthcare, machine learning plays a big role by understanding different parameters and correlate them with diseases. Machine learning is defined as a process in which computers use machine learning algorithms to analyse large set of data that is represented in non-linear ways, identify patterns, and make predictions that can be tested and confirmed.

At a high level, there is supervised machine learning, in which outcomes will be predicted from available data and information from previous outcomes. Further, in unsupervised machine learning, unknown patterns of outcomes will be predicted from data.

Machine learning can be used in healthcare by increasing efficiencies, saving money and saving lives. CT scan data can be analyze and cross-applied to patient records to see who is most at risk

for disease. Physicians can predict post-discharge outcomes to reduce readmissions and enhance patient flow. They can make curative diagnostics faster, more accurate, and more accessible.
Machine learning, deep learning, and cognitive computing are various steps towards a high degree of artificial intelligence, but they are different thing. Deep learning is a subset of machine learning. It uses artificial neural networks that simulates human brain connections that become "trained" with the time to provide answers to questions with nearly 100 percent accuracy.
AI is basically the intelligence that how we make machines intelligent, whereas machine learning is the process of implementing the compute methods that support it. AI algorithms can predict post-discharge results, reducing readmissions and optimizing patient flow. In other words, physicians can use these solutions to make medical diagnostics faster, more precise and manageable. For a caregiver, identifying a pattern is the first step which further solves the complicated process of treatment. Here, artificial intelligence, machine learning will do the same type of thing. Companion diagnostics will help to find the gaps in our current data resources and help us move into truly personalized medicine.

7.3. Anomaly detection to discover unusual patient management
Anomaly detection is used to identify unusual data that diverge from the majority of samples in a dataset. Recently, due to the advancement in data collection and electronic health record, anomaly detection techniques are also helpful in the health care sector. Anomalies are divided into three categories like point anomaly, contextual anomaly and collective anomaly.
To detect an unusual behavior of data, there are many sensors attached to the patient, used to monitor various physiological parameters, and transmit the data to an analytical device which must analyze the collected data, and raise alarms to the physicians only when the patient health degrades.

7.4. Reactive Machine Learning in Healthcare
All types of Healthcare can be divided into two categories: proactive and reactive.

Reactive healthcare includes reacting to an adversarial disease, injury, disorder or symptom. If a patient suffering with a fever and body aches than he may react by visiting the doctor. Depending on the doctor's diagnosis, he or she may recommend you with antibiotics to help to body fight with the infection. Both the patient and the doctor are reacting to the symptoms[29].
Proactive healthcare differs from reactive healthcare in the sense that action need to be taken before the visibility of the symptoms. A patient can take a proactive approach towards his health by boosting the immune system with vitamin C, antioxidants, and by drinking plenty of fluids without waiting to detect the problem.
Healthcare is said to be reactive if it satisfies the four basic principles of reactive system i.e. responsive, resilient, elastic and message driven. Responsive healthcare system refers to proper results of the medicines prescribed by the physicians at all circumstances. Resilient health care system requires the engagement of all the system's components, elastic refers to the healthcare system should be scalable and message driven architecture refers to the request-response messages for exchanging patient details etc. across different hospitals to provide the best of in-Patient and Out-Patient service integration within and across healthcare system.

8. Conclusions
This paper provides a generalized framework for personalized healthcare which influences the advantages of remote monitoring, cloud computing, big data and reactive machine learning. It provides systematic approach to support fast growing data of people with severe diseases. It also simplifies the task of physicians by not overwhelming them with false alerts. The proposed architecture can support for Artificial intelligence bssed healthcare analytics by providing batch and stream computing, extendable storage solution and query management. To achieve more efficient result from healthcare, there is still a need to handle the health care data that is growing time by time at high rate, and larger scale with tons of inconsistent data sources. A distributed system should be arranged to interchange data among labs, hospital systems, and clinical centres. Previous analytics tools have become transparent and friendly. But, when they emerged with open source development tools and platforms, it will become very complex, need complex programming and need the

application of a variety of Refine, KNIME and data mining can be used to improve the efficiency of data analytics.

9. References

[1] R. Weil, '' Big Data In Health: A New Era For Research And Patient Care Alan R. Weil'', Health Affair, Vol. 33, N° 7, pp 1110, 2014.

[2] A. Alexandru, D. Coardos, Big Data in Healthcare and Medical Applications in Romania, IEEE International Conference on Automation, Quality and Testing, Robotics, THETA 20th edition, 2016.

[3] T., Huang, L., Lan, ''Promises and Challenges of Big Data Computing in Health Sciences'', Big Data Research vol. 2, pp 2-11, 2015.

[4] K. Roney, If Interoperability is the Future of Healthcare, What's the Delay?, Becker's Hospital Review, 2012.

[5] Yichuan Wang, LeeAnn Kung, Terry Anthony Byrd, "Big data Analytics: Understanding its capabilities and potential benefits for healthcare organizations", in Press Article," Technology forecasting & social change", Science Direct, Elsevier, 2016.

[6] Jonathan Northover, Brian McVeigh, Sharat Krishnagiri. Healthcare in the cloud: the opportunity and the challenge. MLD.

[7] CSCC, Impact of Cloud Computing on Healthcare, pp. 1-18, 2012.

[8] I.A.T. Hashem, I. Yaqoob, N.B. Anuar, S. Mokhtar, A. Gani, S.U. Khan, The rise of "big data" on cloud computing: Review and open research issues, Information Systems, Vol. 47, pp. 98-115, 2015.

[9] P. Piletic, How is cloud and big data changing healthcare in 2016.

[10] Priyanka Jain, Sanjay Ojha, "Significance of Big Data Analytics", International Journal of Software and Web Sciences (IJSWS), 2015

[11] Basma Boukenze, Hajar Mousannif, Abdelkrim Haqiq, "A Conception of a Predictive Analytics Platform in Healthcare Sector by Using Data Mining Techniques and Hadoop" in Proc. Conf. International Journal of Advanced Research in Computer Science and Software Engineering, Volume 6, Issue 8, August 2016.

[12] Akshay Raul, Atharva Patil, Prem Raheja, Rupali Sawant, "Knowledge Discovery, Analysis and Prediction in Healthcare using Data Mining and Analyics", in Proc. Conference, 2nd International Conference on Next Generation Computing Technologies (NGCT-2016) Dehradun, India 14-16 October 2016.

[13] Frost & Sullivan, Drowning in big data? Reducing information technology complexities and costs for healthcare organizations, 2012.

[14] B. Feldman, E.M. Martin, T. Skotnes, Big Data in Healthcare Hype and Hope, October 2012. Dr. Bonnie 360, 2012.

[15] Akshay Raul; Atharva Patil; Prem Raheja; Rupali Sawant, " Knowledge discovery, analysis and prediction in healthcare using data mining and analytics, in Proc. 2nd International Conference on Next Generation Computing Technologies (NGCT),IEEE, 2016.

[16] M. Roopa, Dr.S. Manju Priya, "A Review of Big Data Analytics in Healthcare" in Proc. Conf. International Journal for Scientific Research & Development| Sp. Issue – Data Mining 2015.

[17] Aditi Bansal and Priyanka Ghare, "Healthcare Data Analysis using Dynamic Slot Allocation in Hadoop" in Proc. Conf. International Journal of Recent Technology and Engineering (IJRTE)

ISSN: 2277-3878, Volume-3 Issue-5, November 2014.

[18] Van-Dai Ta, Chuan-Ming Liu, Goodwill Wandile Nkabinde," Big Data Stream Computing in Healthcare Real-Time Analytics", in Proc. Conf. International Conference on Cloud Computing and Big Data Analysis, IEEE, 2016.

[19] J. Archenaa1 and E.A.Mary Anita, "A Survey of Big Data Analytics in Healthcare and Government" in Proc. Conf. 2nd International Symposium on Big Data and Cloud Computing (ISBCC'15), pp-408 – 413, 2015.

[20] Gualtieri, M; Curran,R, "The Forrester Wave™: Big Data Streaming Analytics, Q1 2016," Forrestor.com, Cambridge MA, 2016.

[21] Wullianallur Raghupathi and Viju Raghupathi, "Big Data analytics in Health care: Promise and Potential", A review paper in Health Information Science and Systems, vol. 2, no. I, Feb. 2014, p. 3.

[22] Rotsnarani Sethy, Mrutyunjaya Panda, "Big Data Analysis using Hadoop: A Survey" in Proc. Conf. International Journal of Advanced Research in Computer Science and Software Engineering, Volume 5, Issue 7, July 2015.

[23] Mukesh Borana 1, Manish Giri 2, Sarang kamble, "Healthcare Data Analysis using Hadoop", in Proc. Conf. International Research Journal of Engineering and Technology (IRJET), Volume: 02 Issue: 07, pp-583, Oct-2015.

[24] nifi.apache.org, "nifi.apache.org/minifi," 18 November 2016.

[25] Dalia AbdulHadi AbdulAmeer, "Medical Data Mining: Health Care Knowledge Discovery Framework Based on Clinical Big Data Analysis" in Proc. Conf. International Journal of Scientific and Research Publications, Volume 5, Issue 7, July 2015.

[26] Meng et al, "MLlib: Machine Learning in Apache Spark," Journal of Machine Learning Research 17, San francisco, 2016.

[27] Telmo da Silva Morais, "Survey on Frameworks for Distributed Computing: Hadoop, Spark and Storm" in Proc. Conf. Doctoral Symposium in Informatics Engineering, 2015.

[28] R. Evans, "Apache Storm, a Hands on Tutorial," IEEE International Conference on Cloud Engineering, Tempe, AZ, 2015, pp. 2-2, 2015.

[29] Harsh Pathak, Manas Rathi, "Introduction to Real-Time Processing in Apache Apex", in Proc. Conf. International Journal of Research in Advent Technology, 2016.

[30] Steve G. Peters, James D. Buntrock,''Big Data and the Electronic Health Record'', Ambulatory Care Manage , Vol. 37, No. 3, pp. 206–210, 2014.

[31] Engineer Bainomugisha, Andoni Lombide Carreton, Tom van Cutsem, Stijn Mostinckx, and Wolfgang de Meuter. A survey on reactive programming. ACM Computing Surveys (CSUR), 45(4):52, 2013.

Local File Disclosure Vulnerability: A Case Study of Public-Sector Web Applications

M. Imran Ahmed[a], Md Maruf Hassan[b], Touhid Bhuyian[c]

Robi Axiata Limited[a], Daffodil International University[b,c]
Dhaka, Bangladesh

Email:maruf.swe@diu.edu.bd

Abstract. Almost all public-sector organisations in Bangladesh now offer online services through web applications, along with the existing channels, in their endeavour to realise the dream of a 'Digital Bangladesh'. Nations across the world have joined the online environment thanks to training and awareness initiatives by their government. File sharing and downloading activities using web applications have now become very common, not only ensuring the easy distribution of different types of files and documents but also enormously reducing the time and effort of users. Although the online services that are being used frequently have made users' life easier, it has increased the risk of exploitation of local file disclosure (LFD) vulnerability in the web applications of different public-sector organisations due to unsecure design and careless coding. This paper analyses the root cause of LFD vulnerability, its exploitation techniques, and its impact on 129 public-sector websites in Bangladesh by examining the use of manual black box testing approach.

Keywords—Web application vulnerability, Local File Disclosure (LFD), public sector.

1. Introduction

Web applications have received huge acceptance and popularity due to rise in the number of internet users. Online features and facilities are not only inspiring users to ask questions and provide support but also creating new opportunities in every business sector. Thus, companies in Bangladesh are increasingly rearranging their work processes and data distribution through web applications for better performance. Digital Bangladesh is one of the major components of Vision 2021. The government of Bangladesh has undertaken several initiatives to implement numerous projects related to digital technology, and a number of those projects are in currently progress such as e-government, e-business, e-learning, e-health, e-employment, e-environment, e-agriculture, e-science, e-citizen services, online banking, online admission, online registration, online ticketing and mobile ticketing [1]. The government had also made it a regulatory requirement for all educational institutions to set up websites within July 2015 [2].

We have observed that a good number web application in the public sector do not meet fully the security requirements in design and coding, which makes these applications very risky in a number of ways. Data/file sharing through the download facility is the most common feature of the website of a public-sector organisation, as it ensures easy distribution of information. Owing to a lack of security of the download process, an intruder can easily exploit the weakness and download files without authorisation, leading to the disclosure of confidential information.

This paper presents an assessment and analysis of the LFD vulnerability with its four major exploitation techniques. We have also tested these four techniques on the websites of different public-sector organisations in Bangladesh. The paper is structured into five sections. Sections 1 and 2 present an introduction to the study and a literature review respectively. Section 3 gives an overview of LFD and a review of its code. After performing a data analysis, results and statistics are presented in Section 4. Section 5 concludes the paper.

2. Literature Review

Many studies have focused on web application vulnerabilities, and its exploitation and prevention techniques. S. Gupta and L. Sharma (2012) used the XAMMP server for experimenting XSS exploitation techniques and implemented the same in blogs. In their review, they found that the sandbox environment of web browsers would be a novel technique for mitigating XSS vulnerability [3]. S. Chavan and D. Meshram (2013), who identified the root cause of different web application vulnerabilities, segregated the problems from the phase of SDLC. They also proposed countermeasures for those vulnerabilities [4]. R. Johari and P. Sharma (2012) presented a detailed review of SQLi and XSS vulnerability attacks, and the techniques to prevent them [5]. O. B. Al-Khurafi et al. (2015) conducted a survey on SQLi, XSS, broken authentication and session management, and described different attack techniques of the above-mentioned vulnerabilities [6].

Several researches also studied different web application vulnerabilities such as Structured Query Language Injection (SQLi), Cross Site Scripting (XSS), broken authentication, Cross Site Request Forgery (CSRF), Local File Inclusion (LFI), and Remote File Inclusion (RFI), along with their exploitation of Bangladeshi domains. D. Alam et al. (2015) investigated the user-input-based SQLi technique implemented on the web applications of .bd domain [7]. They also studied three major SQLi techniques implemented on Bangladeshi educational websites and analysed the impact after exploitation [8]. T. Farah et al. (2015) conducted a similar study on financial web applications in Bangladesh and provided the security status. T. Farah et al. (2016) also studied XSS and CSRF vulnerabilities of Bangladeshi web applications. In their article, the researchers drew the attention to coding flaws in the web applications and prescribed some solutions [9]. A. Begum et al. (2016) provided a detailed description of LFI vulnerability exploitation techniques based on RFI and SQLi, and showed the impact after conducting the given exploitation on 153 web applications of Bangladesh [10]. M. Hassan et al. (2016) conducted a case study of LFD exploitation on educational websites in Bangladesh [11].

The above literature review shows that there has been no examination of LFD exploitation on public-sector web applications in Bangladesh. In this paper, we will investigate the public-sector web applications using four types of LFD exploitation techniques.

3. Overview of LFD and Its Code Review

LFD is a kind of vulnerability that allows users to download sensitive files/information from a website in an unauthorised manner by misusing download features of the application. For example, on a web application which is built on PHP and where LFD vulnerability is present, users may access several restricted files, e.g. config.php/, boot.ini/, and index.php, using directory traversal techniques, by changing the file path of infected parameters. By downloading those secret files, users may get their hands on very confidential information like usernames, passwords, database username, and port numbers.

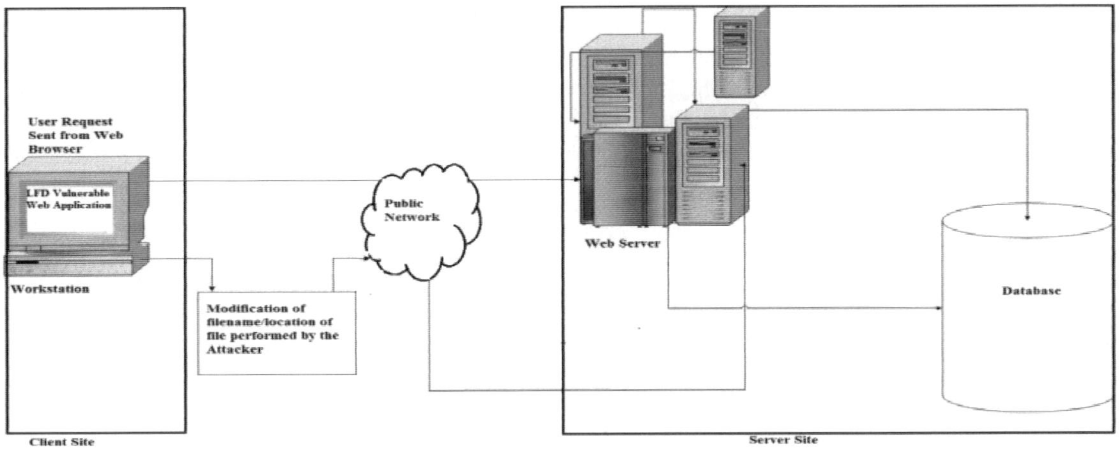

Figure 1: LFD Vulnerability Exploitation [11]

A web application comprises two parts, namely the server side and the client side. As soon as a user of the application makes a request for a service (e.g. downloading a file, request for information, etc.) from the web browser, i.e. client side, it refers the request to the main web application server. From the server side, the application will process the requested or retrieved information from the database and return it to the client-side browser's interface. Figure 1 shows how a hacker can exploit the LFD vulnerability [11].

3.1. Cause of LFD Vulnerability

An inappropriate use of some built-in functions of popular PHP programming language, e.g. read_file(), file(), and file_get_contents(), causes LFD vulnerability.

```
Line 01: <?php
Line 02: $vulnerable_file=readfile($_GET['file_name']);
Line 03: Print_r($vulnerable_file);
Line 04: ?>
```

Figure 2: Example of LFD-vulnerable coding pattern using $_GET method

Figure 2 above gives an example of PHP code with LFD vulnerability. The code highlights the areas of mistake that might occur during the development period leading to LFD vulnerability. The function readfile() is used to read a specific file from a particular location on the host server (Line 02, Figure 2). However, the use of the function results in the LFD vulnerability due to the function not being closed, causing serious loopholes in the code. In this case, a user will be able to access files illegally by changing the specific filename/location.

3.2. Exploitation Techniques

There are four types of LFD exploitation techniques:

3.2.1 General LFD exploitation
3.2.2 Base 64/ URL/Hex encoding exploitation
3.2.3 Long directory transversal exploitation
3.2.4 HTTP POST request base exploitation

3.2.1. General LFD Exploitation. LFD vulnerability is normally found in the document sharing sections of web applications. By simply clicking on the file download button, a user can view/download the specific shared file (e.g. document.pdf). After that, the file_get_content() function connects to the server's database and allows the user to send his query to the database for to access the selected file. In this case, the download operation is conducted using the $_GET method, in which the file parameter/name is displayed on the URL bar. If a user changes the file name, the given file will be downloaded straight way in case LFD vulnerability exists in the application. Consider a web application having the following URL with anyfile.pdf to download:

http://www.demoname.com/files/shared.php?file=anyfile.pdf

The above URL shows a page with a file, anyfile.pdf, ready to be downloaded. If the file name is modified by the user, the downloading process will continue in case the said filename is present in the specific directory. For example, if the filename of the above URL is changed as follows, it is clear that config.php will be downloaded.

http://www.demoname.com/files/shared.php?file=config.php

Once the user can download the configuration file, the following web application server information ($hostname = 'specific server'; $username = 'root'; $password = 'dbpass'; $database = 'dbname'; $connector = mysql_connect ($hostname, $username, $password); mysql_select_db) can be disclosed as well. Using the same technique, one can even download the source code of the index.php file, which is very alarming for the host of the web application.

3.2.2. Base 64/ URL/Hex Encoding Exploitation. To hide the filename from the web application's URL, developers use encoding methods, e.g. base64, hex, and URL encoding, for security reasons.

http://www.demoname.com/files/shared.php?file =YW55ZmlsZS5wZGY=

Base64 encoded filename i.e. document.txt

The above URL is an example where the filename is encoded using base64. To decode the URL, the attacker first has to identify the encoding methods and reveal the filename using online decoding tools, in this case base64. The attacker then sends the request to the server with the URL, along with the decoded filename. Using this technique, an unauthorised user can download confidential files.

3.2.3. Long Directory Traversal Exploitation. For security reasons, developers sometimes store files in different directories, thereby making it difficult for the attacker to find the desired file. In this case, the attacker uses directory traversal query for accessing the desired file to bypass this security trick. By adding command (../), the user can get access to the previous directory.

http://www.vulnsite.com/downloads.php?file=../../../../../../../config.php

Directory changing process

If the above config.php does not appear in the given directory where the downloading file is located, every time the '../' command is used, it will go to the next directory. It will explore the root and check whether the desired file exists or not. The file in question will automatically download if and when the intruder selects the right directory.

3.2.4. Long Directory Traversal Exploitation. Web developers usually design data processing techniques through HTTP POST methods for enhanced security. A general user will not be able to view the transformation data easily as it is stored in the cookie. An attacker can bypass this restriction if the code contains a design bug as follows:

```
/* code start
Line 01: $var_namee = readfile($_POST['anyfile']);
/vulnerable coding pattern ($_POSTmethod)
Line 02: Print_r($vulnerable file);
Line 03: ?>
 */ code end
```

Figure 3: Example of LFD-vulnerable coding pattern using $_POST method

In the above code in Figure 3, the data parameter would not be seen on the URL bar as it uses the POST method. Using tools like live HTTP headers, Burp Suite, temper data, and the regular HackBar on Firefox, a user can retrieve cookies where the download file information can be visible and the filename can also be modifiable.

3.3. Test Case

The test cases stated below are being followed to conduct our study for exploiting LFD vulnerable public sector web applications of Bangladesh.

i. Check whether the file name is visible in the URL or not after clicking the downloading link of the web application.
ii. Check whether the cookies can be retrieved or not in case the file name is not visible in the URL.
iii. Check whether the filename is in plaintext or in encoded format.
iv. Check whether the encoded filename can be converted into plaintext or not using tools.
v. Check whether the filename can be modified from the URL or not.
vi. Check whether the desired file can be discovered through directory traversal technique or not in case the file is not in public download folder of the web application.
vii. Check whether the concerned file is downloaded or not after modifying the desired file name in URL.
viii. Check whether the database connectivity can be established using third party software or not after getting the respective information from the downloaded confidential file.

4. Result Analysis

In this research, we have articulated the sample size mechanism using the universal calculator provided by G*Power 3.1.9.2. We studied 129 websites of public-sector organisations in Bangladesh using the random sampling method. Of these, 29.45% websites were found to have LFD vulnerability. The existence of four types LFD vulnerability was found on those websites. We used the manual black box testing approach to collect data for this study. We analysed this data set according to the LFD exploitation type, the institution type, and the level of access to the host system. The analysis is discussed below:

4.1. Result Based on the Access Level

Figure 4 represents the level of access to the host system achieved after exploiting the LFD vulnerability of 38 public-sector web applications.

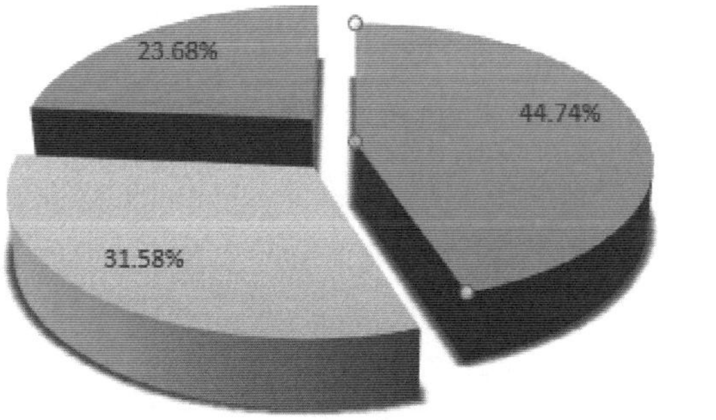

Access Level	Percentage
Full Access obtained	44.74%
Accessible to the database	31.58%
System parameter disclosure	23.68%

Figure 4: Level of Access in Host System

In this examination, full access to the host system was obtained in the case of 44.74% websites, whereas only database access could be established on 31.58% websites through remote access using third-party software. In our case, we used HeidiSQL for connecting with the host's database system remotely. On 23.68% of the sites, system parameters such as port number, username, password, and IP address could be accessed; however, we were unable to access such information due to a port forwarding problem.

4.2. Result Based on the Institution Type

Figure 5 shows institution-wise LFD exploitable web applications of Bangladeshi public-sector organisations. The figure shows that LFD vulnerability exists mostly on the web applications of government organisations and public universities in Bangladesh, their respective percentages being 21.05% and 28.95%.

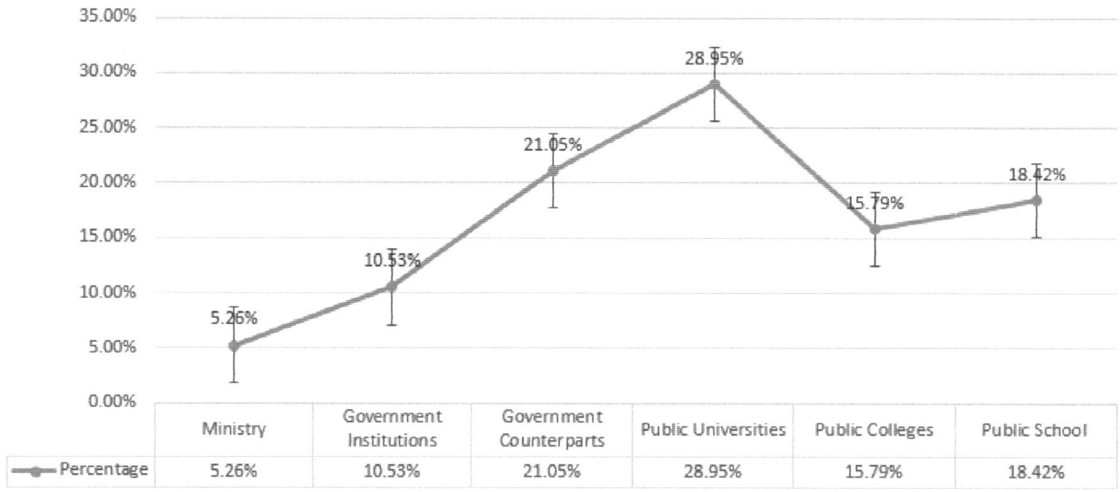

Figure 5: Public Institution wise LFD vulnerability exploitation

The same weakness was found on the websites of 15.79% of public colleges and 18.42% of state-run schools. The web applications of 10.53% of government institutions have this vulnerability. Noticeably, only 5.26% of ministry websites suffer from the above weakness.

4.3. Result Based on LFD Exploitation Types

Figure 6 gives the percentages of Bangladeshi public-sector web applications that are exploited by LFD vulnerability exploitation techniques.

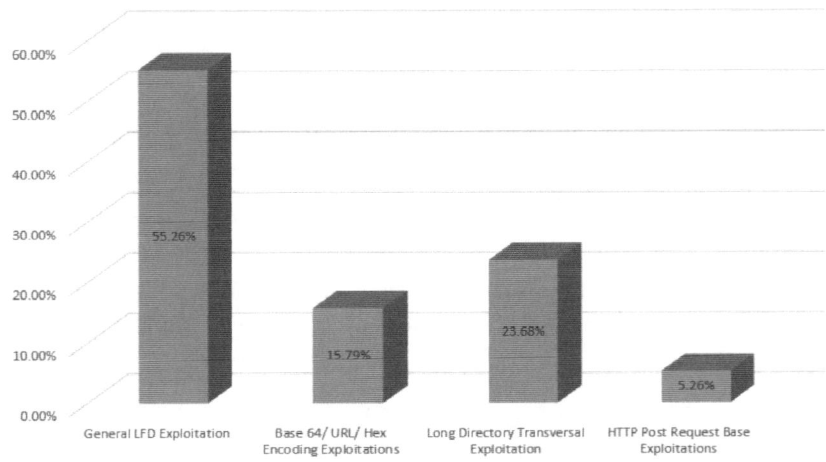

LFD Type	Percentage
General LFD Exploitation	55.26%
Base 64/ URL/Hex Encoding Exploitation	15.79%
Long Directory Transversal Exploitation	23.68%
HTTP Post Request Base Exploitation	5.26%

Figure 6: LFD Exploitation Types

Figure 6 gives the percentages of Bangladeshi public-sector web applications that are exploited by LFD vulnerability exploitation techniques. It can be seen that more than half the LFD-vulnerable sites in our sample are exploitable through General LFD Exploitation. HTTP Post Request Base Exploitation is the least-used technique to get unauthorised information/access. Long Directory Transversal Exploitation and Base 64/URL/Hex Encoding Exploitation are used 23.68% and 15.79% respectively.

5. Conclusion

Owing to their versatility and easy accessibility, the web applications of public-sector organisations in Bangladesh are growing in importance. In our evaluation of 129 such websites, we found that web applications with LFD vulnerability would concede their host's administrative access to unauthorised intruders in most cases. This vulnerability would give an attacker the ability to manipulate sophisticated and confidential documents such as student selection lists for admission and academic results. Many websites of public universities in Bangladesh contain LFD vulnerability, which makes their system very unsafe due to the risk of the disclosure of sensitive data. This may have adverse implications not only for the future of the students, but also for the reputation of the university. Our research in this area is a continuous process. Our observation reveals that the presence of LFD vulnerability in our domain of study is created only because of the carelessness of application developers. Errors on the developers' part include, for example, not closing the readfile() function, not using the SWITCH function, etc. We believe that careful coding and regular monitoring during the development of web applications will mitigate the risk of such oversight that causes serious vulnerability in the applications.

5.1. Limitation & Future Work

We conducted our study using manual black box testing and did a review of public-sector web applications. In the future, we will design an LFD detection model and implement it using automated detection tools.

References

[1] "Digital Bangladesh: Dreams and reality", 2017. [Online]. Available: http://www.thedailystar.net/supplements/24th-anniversary-the-daily-star-part-1/digital-bangladesh-dreams-and-reality-73118

[2] "School & college compulsory website of Bangladesh", 2017. [Online]. Available: http://allnewsview.com/school-college-compulsory-website-of-bangladesh [Accessed: 11 September 2016]

[3] Gupta, S., & Sharma, L. (2012). Exploitation of cross-site scripting (XSS) vulnerability on real-world web applications and its defense. International Journal of Computer Applications, 60(14).

[4] Chavan, S. B., & Meshram, B. B. (2013). Classification of web application vulnerabilities. International Journal of Engineering Science and Innovative Technology (IJESIT), 2, pp. 226–34.

[5] Johari, R., & Sharma, P. (2012, May). A survey on web application vulnerabilities (SQLIA, XSS) exploitation and security engine for SQL injection. In Communication Systems and Network Technologies (CSNT), 2012 International Conference on (pp. 453–58). IEEE.

[6] Al-Khurafi, O. B., & Al-Ahmad, M. A. (2015, December). Survey of web application vulnerability attacks. In Advanced Computer Science Applications and Technologies (ACSAT), 2015 4th International Conference on (pp. 154–158). IEEE.

[7] Alam, D., Kabir, M. A., Bhuiyan, T., & Farah, T. (2015, October). A case study of SQL injection vulnerabilities assessment of .bd domain web applications. In Cyber Security, Cyber Warfare, and Digital Forensic (CyberSec), 2015 Fourth International Conference (pp. 73–77). IEEE.

[8] Alam, D., Bhuiyan, T., Kabir, M. A., & Farah, T. (2015, November). SQLi vulnerabilty in education sector websites of Bangladesh. In Information Security and Cyber Forensics (InfoSec), 2015 Second International Conference on (pp. 152–57). IEEE.

[9] Farah, T., Shojol, M., Hassan, M., & Alam, D. (2016, July). Assessment of vulnerabilities of web applications of Bangladesh: A case study of XSS & CSRF. In Digital Information and Communication Technology and its Applications (DICTAP), 2016 Sixth International Conference (pp. 74–78). IEEE.

[10] Begum, A., Hassan, M. M., Bhuiyan, T., & Sharif, M. H. (2016, December). RFI and SQLi based local file inclusion vulnerabilities in web applications of Bangladesh. In Computational Intelligence (IWCI), International Workshop on (pp. 21–25). IEEE.

[11] M. M. Hassan, T. Bhuiyan and S. Biswas, "An Investigation of Educational Web Applications in Bangladesh: A Case Study on Local File Disclosure Vulnerability", 2016 4th International Conference on "Engineering & Technology, Computer, Basic & Applied Sciences" (ECBA-2016), Sydney, Australia, (pp-11-16).

Acknowledgments

We thank the authorities of the organisations who have given us permission to examine their websites.

IOP Conf. Series: Journal of Physics: Conf. Series **933** (2017) 012012 doi:10.1088/1742-6596/933/1/012012

Spam comments prediction using stacking with ensemble learning

Arif Mehmood[1], Byung-Won On[2], Ingyu Lee[3], Imran Ashraf[1] ,Gyu Sang Choi[1]

[1]Department of Information and Communication Engineering, Yeungnam University, Gyeongbuk, Republic of Korea
[2]Department of Software Convergence Engineering, Kunsan National University, Republic of Korea
[3]Sorrel College of Business, Troy University, Troy, AL 36082, USA

Corresponding authors:castchoi@ynu.ac.kr[1] , on.byung.won@gmail.com[2]

Abstract. Illusive comments of product or services are misleading for people in decision making. The current methodologies to predict deceptive comments are concerned for feature designing with single training model. Indigenous features have ability to show some linguistic phenomena but are hard to reveal the latent semantic meaning of the comments. We propose a prediction model on general features of documents using stacking with ensemble learning. Term Frequency/Inverse Document Frequency (TF/IDF) features are inputs to stacking of Random Forest and Gradient Boosted Trees and the outputs of the base learners are encapsulated with decision tree to make final training of the model. The results exhibits that our approach gives the accuracy of 92.19% which outperform the state-of-the-art method.

1. Introduction

The Spam in current era of digital communication is a trick for diverting the traffic.The continuous growth of users on social networkslike Facebook, Twitter, YouTube, etc., is also a reason of massive information spread. This information usually spread through reviews and comments that have opened new avenues for spammers.The famous video social websites like YouTube, Daily motion and Vimeo are also infested with spam that normally embroils comments and links to some salacious or dating site or some irrelevant videos. These comments are usually generated automatically through bots software.Predicting spam information is a challenge for the researcher in the field of natural language processing [1].Typically, it is initiated for getting the people's attention through deceptive comments and reviews [2].

In this study, we address above mentioned challenges and propose a novel spam comments prediction model based on a stacking with ensemble learning on text features of TF/IDF [3]. In comparison to other methods which are single learning classifiers our proposed model is based on combining the classifier functionality to achieve the maximum accuracy in prediction. We apply stacking that is a mechanism in which the output of the classifiers at level one will be used as training data for another classifier to approximate the same target function. At level one, we apply Random Forest (RF) [4] and Gradient Boosting Tree (GBT) [5] and for level two we use simple Decision Tree (DT).

The proposed model has been evaluated on data sets of UCI Machine Learning Repository [6] about the YouTube Spam Collection. Results of experiments show that our proposed model significantly outperforms the state-of-the-art methods.
The rest of the paper is organized in the following gmanner. Section 2 presents related work. Section 3 gives details of our proposed stacking model for spam comment prediction. Section 4reports experimental results and comparison with other models and section 5 concludes this work. In the end future works in this domain is highlighted.

2. Related work

Spam data prediction is one of the key security challenges in the field of social networks as well as cyber world in general [7] [8].Usually for detection and filtrations of unwanted information blacklists are used e.g. Twitter uses the blacklists filtering module BotMakerin anti-spam system [9].Social media spam is a type of spam information which spreads rumors on the social media [10].

Freedom of free comment in social media is largely misused by spammers especially in YouTube Sureka [11] proposes a data mining based methodology to predict spam comments in YouTube forums. Radulescu, Dinsoreanu, and Potolea [12] proposes a methodology based on various features such as discontinuous text, inadequate and vulgar content and unrelated and out-of-context comment. They mainly focus on the features extraction but our approach is using simple text features TF/IDF and most importantly the combining of the ensemble learning. Alberto, Lochter, and Almeida [13] introduces an online tool called TubeSpam for automatically detecting spam comments posted in the comments sections of YouTube videos. The authors also evaluates different state-of-the-art classification techniques and conclude that decision trees, logistic regression, Bernoulli naïve Bayes, random forests, and support vector machines are statistically equivalent. In our proposed methodology, we uses different learning technique that not been used previously.

3. Proposed methodology

Our methodology is based on multi learner models and usually it is called stacking. The selections of models at the stage of base learners and at accumulate stage is novel to the best of our knowledge. The flow diagram of proposed methodology is shown in Fig. 1.The proposed methodology has been tested on YouTube spam comments data sets at UCI repository. Data set includes 1956 spam and ham documents that are further divided in the ratio of 70% for training and 30% for test purpose. For training we have total of 1370 documents that include 666 ham and 301 spam. Similarly, division for test we have 586 that include 285 ham and 301 spam documents.

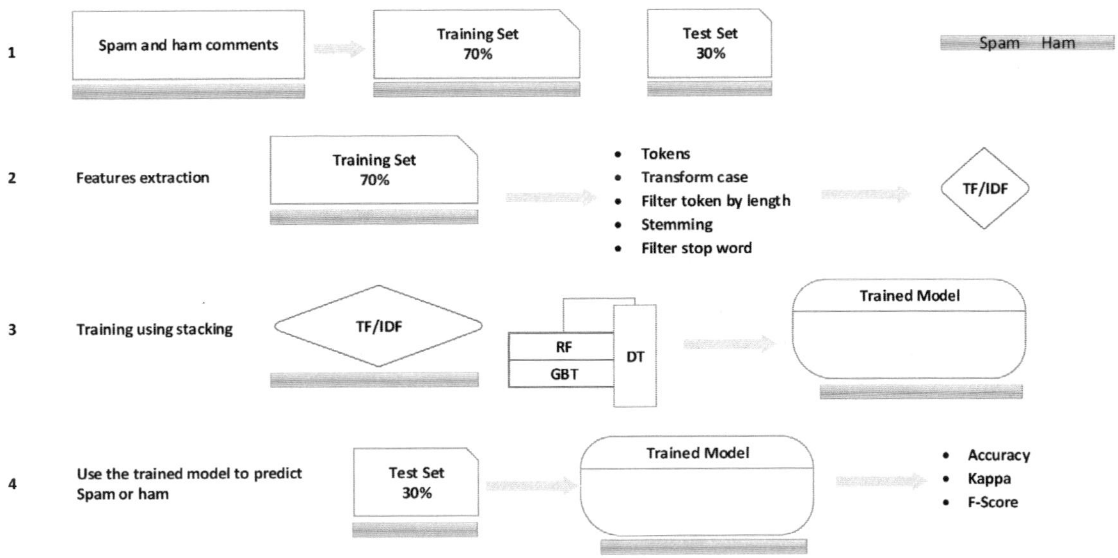

Figure 1. Proposed methodology

We use RF and GBT at base learners. Both models are ensemble learning models and arelater combined with the DT model. We extract only TF/IDF text features as described in (1) for training purpose. The raw spam text has been filtered through different process like tokenization, transform case, stemming, filter token by length and applying the stop words. Porter stemming algorithm[14] is used for stemming and filter token with a length of minimum of 04 characters and maximum of 25 characters.

$$W_{i,j} = tf_{i,j} \times log(\frac{N}{df_i}) \quad (1)$$

Where term frequency i.e., $tf_{i,j}$ is the number of occurrences of i in j and df_i is the number of documents containing i while N is total number of documents.

4. Results and Discussion

We choose the best of ensemble learning models at stage one and they cover the deficiency of each other [15].We keep the TF/IDF features for training and testing the proposed methodology. The same features also tested on various single state-of-the-art learning model and same combination as our proposed approach but Naïve Base (NB) at stage2 in stacking. However, we find that the proposed methodology outperforms other tested models.

The results are shown in Table 1. GBT also shows good results working alone and performs well with RF. However, at stage 2 it does not show good results when working with NB and its performance is not good enough to meet our scheme. The other models like RF and Support Vector Machine (SVM) do not perform well when using the TF/IDF features. The reason our strategy outperforms is that both RF and GBT are ensemble learner based on DT. In addition, we also use DT at stage 2 in stacking which improves the results.

Table 1. Different models accuracy comparison

Methodology	Accuracy
GBT	90.66%
RF	83.02%
SVM	81.49%
(GBT+RF)-NB	91.00%
Proposed	92.19%

The percentage of accuracy, kappa value and F-scores are used as the performance parameters to evaluate the accuracy of all classifiers. Fig. 2 exhibits that the performance of the proposed methodology is significantly better than the other selected models.

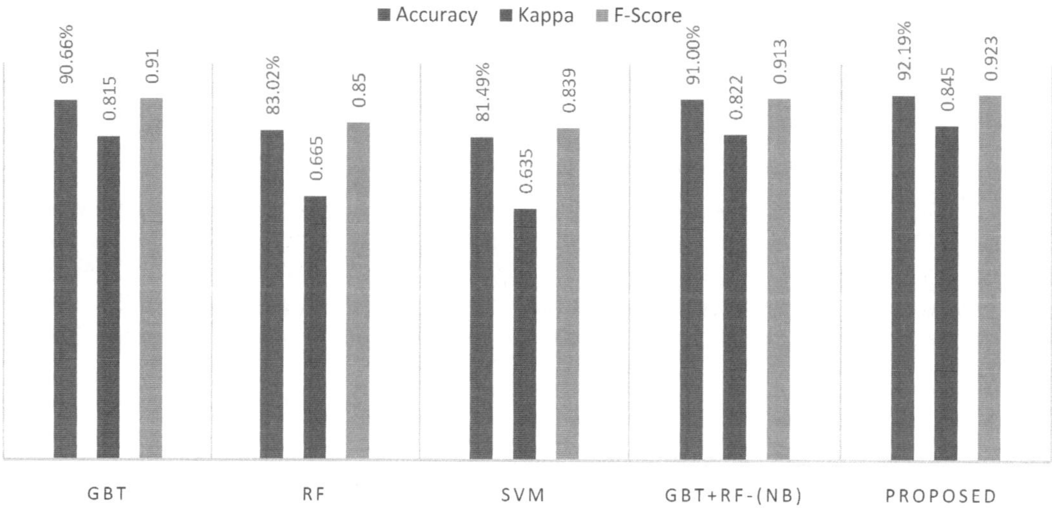

Figure 2. Performance comparison

The percentage of accuracy, kappa value and F-scores are used as the performance parameters to evaluate the accuracy of all classifiers. Fig. 2 exhibits that the performance of the proposed methodology is significantly better than the other selected models. Confusion matrix is shown in Table 2 that exhibits the strong agreement on both ham and spam. The performance of the model in terms of precession and recall are 0.888 and 0.972 respectively.

Table 2. Confusion Matrix of the proposed model

Confusion Matrix	Ham	Spam
Ham	278	35
Spam	8	265

5. Conclusion and future work

This paper proposes a new strategy for spam comments prediction using simple but famous text features of TF/IDF for novel selection of stacking model at base learner and final learning. This

strategy performance gives the accuracy of 92.19% that is significantly better as comparedto single model base or even stacking with NB.
We are planning to test the model on different spam data sets to evaluate the efficiency of the model in the future.

6. References

[1] F. Figueiredo, J.M. Almeida, M.A. Gonalves, F. Benevenuto.Trendlearner: early prediction of popularity trends of user generated content. Inf. Sci. (Ny), 349 (2016), pp. 172–187.
[2] D. Streitfeld.For 2 a Star, an Online Retailer Gets 5 Star Product Reviews26, , New York Times (2012)
[3] G. Salton and C.Buckley, 1988. Term-weighting approaches in automatic text retrieval. Information processing & management, 24(5), pp.513-523.
[4] L. Breiman,Random forests. Mob. Learn. 2001, 45, 5–32.
[5] J.H. Friedman, Greedy function approximation: A gradient boosting machine. Ann. Stat. 2001, 29, 1189–1232.
[6] UCI YouTube Spam Collection Data Set https://archive.ics.uci.edu/ml/datasets/YouTube+Spam+Collection(accessed on 01 July 2017).
[7] F. Norouzi, A. Dehghantanha, B. Eterovic-Soric, K.-K.R. Choo. Investigating social net-working applications on smartphones: detecting Facebook, Twitter,LinkedIn, and Google+ Artifacts ON Android and iOS platforms Aust J Forensic Sci (2015), 10.1080/00450618.2015.1066854
[8] D. Quick, B. Martini, K. Choo Cloud storage forensics Syngress Publishing, Waltham (MA) (2014)
[9] R. Jeyaraman Fighting spam with botmaker Twitter Engineering Blog, August (2014)
[10] F. Wu, J. Shu, Y. Huang, Z. Yuan Co-detecting social spammers and spam messages in microblogging via exploiting social contexts Neurocomputing, 201 (2016), pp. 51–C65
[11] A. Sureka Mining user comment activity for detecting forum spammers in youtube CoRR, abs/1103.5044 (2011)
[12] C. Radulescu, M. Dinsoreanu, R. Potolea Identification of spam comments using natural language processing techniques Intelligent computer communication and processing (iccp), 2014 ieee international conference on, IEEE (2014), pp. 29-35
[13] T.C. Alberto, J.V. Lochter, T.A. Almeida Post or block? Advances in automatically filtering undesired comments Journal of Intelligent & Robotic Systems, 80 (2015), pp. 245–259 https://doi.org/10.1007/s10846-014-0105-y
[14] M. Porter, Porter Stemming Algorithm. Available online: http://tartarus.org/martin/PorterStemmer/ (accessed on 01 July 2017).
[15] A. Mehmood,B.W.On, I. Lee, and G.S. Choi,., 2017. Prognosis Essay Scoring and Article Relevancy Using Multi-Text Features and Machine Learning. Symmetry, 9(1), p.11.

Acknowledgments
This research was supported by the National Research Foundation of Korea (NRF) grant funded by the Korea government (MEST) (No. NRF-2016R1A2B1014843) for the second author (Byung-Won On), and supported by the Ministry of Trade, Industry & Energy (MOTIE, Korea) under Industrial Technology Innovation Program. No.10063130, by Basic Science Research Program through the National Research Foundation of Korea (NRF) funded by the Ministry of Education (2016R1A2B4007498), and the MSIP(Ministry of Science, ICT and Future Planning), Korea, under the ITRC(Information Technology Research Center) support program (IITP-2017-2016-0-00313) supervised by the IITP (Institute for Information & communications Technology Promotion) for the fifth author (Gyu Sang Choi).

IOP Conf. Series: Journal of Physics: Conf. Series **933** (2017) 012013 doi:10.1088/1742-6596/933/1/012013

Data Leakage Prevention: E-Mail Protection via Gateway

[1]Kamaljeet Kaur, [2]Ishu Gupta and [3]Ashutosh Kumar Singh

[1]Govt. Sr. Sec. School, Ambala, Haryana, India

[2, 3]Department of Computer Applications, National Institute of Technology, Kurukshetra, Haryana, India

er.kamal1986@gmail.com, ishugupta23@gmail.com

Abstract. Protection of digital assets and intellectual property is becoming a challenge for most of the companies. Due to increasing availability of database services on internet, data may be insecure after passing through precarious networks. To protect intellectual property (IP) is a major concern for today's organizations, because a leakage that compromises IP means, sensitive information of a company is in the hands of biggest competitors. Electronic information processing and communication is replacing paper in many applications increasingly. Instead of paper, an email is being used for communication at workplace and from social media logins to bank accounts. Nowadays an email is becoming a mainstream business tool. An email can be misused to leave company's sensitive data open to compromise. So, it may be of little surprise that attacks on emails are common. So, here we need an email protection system (EPS) that will protect information to leave organization via mail. In this paper, we developed an algorithm that will offer email protection via gateway during data transfer. This algorithm matches the patterns with the keywords stored in the database and then takes the actions accordingly to protect the data. This paper describes why email protection is important? How companies can protect their confidential information from being leaked by insiders.

1. Introduction

In information security, data leakage is unwanted disclosure of confidential information. Companies placed important data like Social Security no's, credit card information, data related to their financial property etc. in their databases; therefore protection of data must have the first priority [1] [2]. Data security is comprised through unintentional and unwise behaviors of employees and IT professionals. The survey shows that employees around the world are responsible to put corporate & personal data at risk [3]. When an authorized user of company discloses confidential information to external entities called as a data leakage.

Data leakage has great impact on the company's reputation [4]. According to survey report done by IBM, 46% of the companies suffered from reputational damage due to data leakage, including confidential employee information, customer information etc. Data leakage incidents usually occur when employees performed everyday tasks, such as sending e-mail that inadvertently contains sensitive information [5]. The main reason for using email is possibly the convenience and the speed with which it can be transmitted, irrespective of geographical distances. Sensitive information, such as account statements, credit card scores, and information about sanctions, is being exchanged through emails [6]. According to industry analysts, email volume in companies is growing by more than 30% annually, and the average user receives 7MB of data per day via email. As a result of this growth, handling an email is becoming a critical business.

As shown in Figure 1, three main principles of Information Security are involved in maintaining the *Confidentiality, Integrity*, and *Availability* of information resources. A fault in any one of these three key areas will weaken the security posture of an email system and open the door to misuse. Therefore, the contents of emails are more important and valuable than ever, and their security has raised many concerns [7].

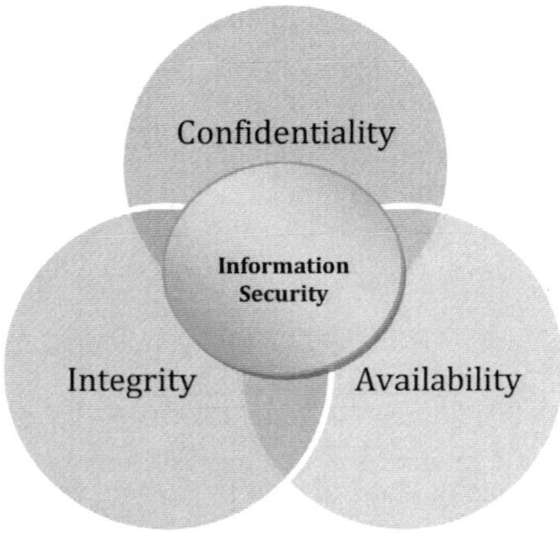

Figure 1: Three main principles of security.

2. Threats to an E-Mail

Employees put company's sensitive data and assets at risk despite policies that define correct procedures. The following examples show how employees intentionally and unintentionally leak sensitive data.

- *Usage of Unauthorized applications*: In companies, usage of personal emails can place sensitive data and personal information at risk. According to a survey report, 63 percent of employees admit that they use work computer for personal usage. These applications do not follow corporate security standards. As a result, data leakage by an employee is high.

- *Misuse of Corporate Computers*: Employees intentionally use company's computers in many ways that undermine IT security policies which includes sharing of work devices and sensitive information with non-employees. These behaviors can result in leaking out the IP of company, which poses serious threats to company's security and profitability.

- *Misuse of Passwords and Login/Logout Procedures*: When an employee leaves a system logged in and with a password attached to it that invites an attacker to steal the sensitive data at their leisure. If employee used that computer for personal use which means information is now willingly available to the attacker.

3. Why E-Mail Protection is Important?

Email is used by millions of people to communicate around the globe and is a critical application for many businesses. Email messages passes through intermediate computers before reaching their final destination and it is relatively easy for attackers to intercept and read messages. The backups of these can remain up to several months on their server, even if we delete them from our mailbox.

Employee of an organization accesses important files and sends it by their personal mails or by any other web service to the outside of the network. A total of 95% of such losses are unintentional. In fact, in 2015, over 200 billion emails were sent & received each day worldwide [8] [9].

➢ **E-Mail Protection via Gateway**

Data leakage prevention solution can be configured on three levels *i) Individual user level ii) Domain level iii) At gateway level.*

- *Individual user level:* Some companies facilitates to their employees that they can complete their work from home, in this situation their personal computers should be connected to their office PC via RSA token.
- *Domain level:* A company can prevent data leakage on domain level by enforcing the rule that an employee must be restricted to the company's domain. But still there are departments in companies that have rights to send information outside of the network like the sales department of the company.
- *Gateway level:* In a work place, gateway computer route all the traffic from one network to another. When an employee of the company sends an email to the outside of the network, firstly mail passes through gateway computer then it leaves the network [10] [11].

4. Proposed System

Usage of email is increasing day by day and it is vulnerable in many ways. Security of an email is the main concern for companies & it includes confidentiality that ensures information will not expose to unauthorized entities.

To maintain confidentiality and integrity of an email, we proposed an Email Protection System. Our proposed EPS consider two parameters i.e. data states and deployment. In general data consist three states i.e. data at rest, data in use and data in motion. In this paper our focus is on the state when data moves from one network to another. So we deployed our Email Protection System on gateway that will protect an email consisting confidential information.

As shown in Figure 2, in companies there are sensitive departments that consists sensitive information and they have right to send any information outside of the network for example: Sales department of the company. To keep information safe, administrator enforces an Email Protection policy by selecting important keywords of the documents into a table stored in data base. Our EPS will automatically check all outgoing mails from the sales department and take action based on what the administrator has specified in the policy.

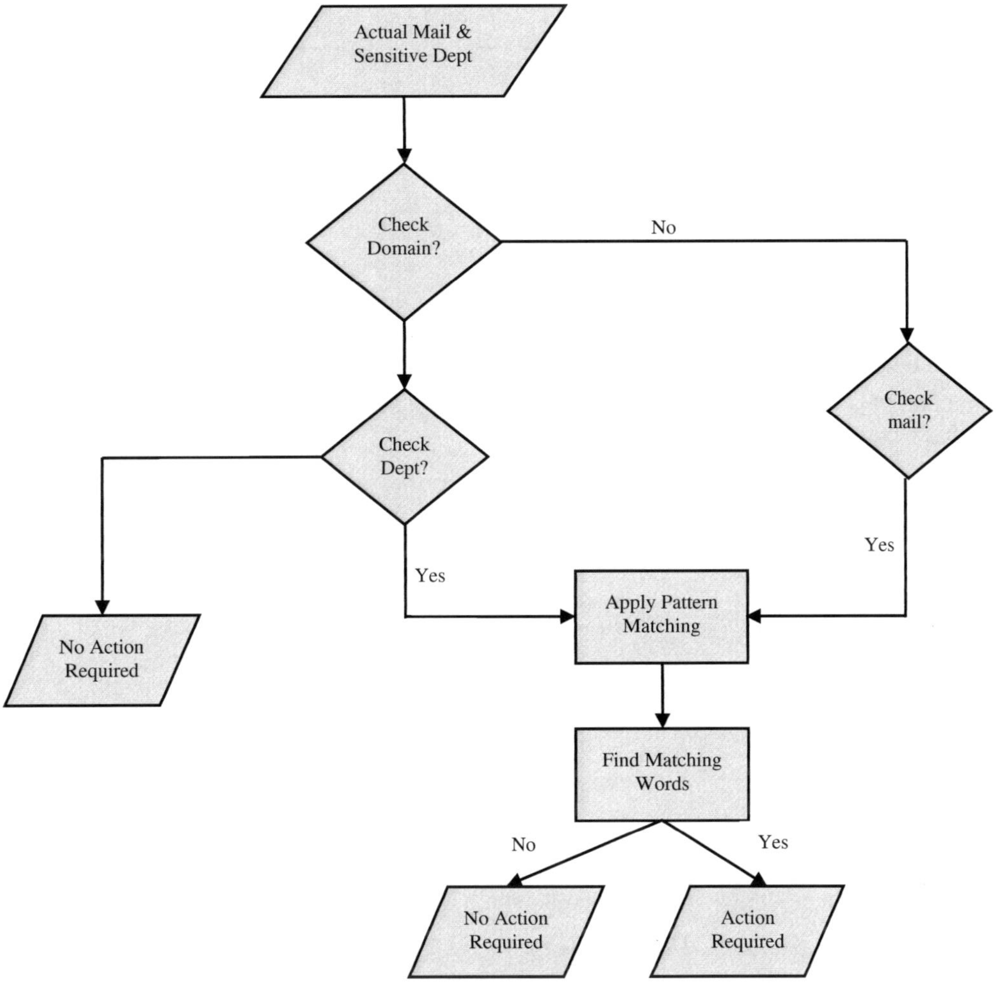

Figure 2. Email protection system deployed on gateway level

5. Proposed Algorithm

EPS checks the actual mail (including attachment) that is ready to leave the network, whether it is sending in our domain or outside of the network. After finishing, it checks the department, i.e. which department is sending an Email like Sales department or any other. Sales department creates a spreadsheet with name, credit card information, and social security no's and prepares it monthly. Department attach this spreadsheet into an email and sends it to a business partner

EPS opens the email consisting spreadsheet inside it which is created by the sales department. Protection System read it through, checks all of the cells of spreadsheet and apply the policy that administrator put in place of matching keywords. EPS not only scans the subject and copy but also scan the content inside it. EPS will match all the keywords of Spreadsheet with the table having important keywords of sensitive documents that are stored in company's database by using PATTREN-MATCHING ALGORITHM. After finding matching words, administrator can take appropriate action.

Algorithm: Pattern-Matching Algorithm

Input: T1, T2 (Pattern), Dept, and Mail
//T2: Master Pattern Table
//Dept: sensitive department like sales department of company
Output: J, the Pattern match join table of T1 and T2

```
begin
J = 0;
If match(Mail.To, Company Domain) AND not match(Mail.dept, Dept) then
  Return J;
end if
If match((Mail.To, Company Domain) AND match(Mail.dept, Dept)) OR not match(Mail.To,
Company Domain) then
  for all t1∋ Mail do
    for all t2 ∋ T2 do
      If match (t1, t2) then
        J = J Join (t1 ↔ t2);
      end if
    end for
    If J ≠ 0 then
      Apply Action ( );
    end if
  Return J;
  end for
end if
end
```

6. Necessary Actions for an E-Mail Protection

After finding anything malicious, administrator can take necessary actions which are as follows:

- *Block an email:* Once scanning is complete, extracted keywords of an outgoing mail are matched with the keywords stored in the database, the administrator can block an Email if found anything malicious.

- *Encrypt an Email:* To protect sensitive information, the administrator can encrypt an email by using some encryption algorithms.

- *Quarantined an Email*: On finding suspicious message or attachment, administrator has right to isolate it apart from the mailbox so that appropriate measures can be taken to read or retrieve it.

7. Conclusion

Data leakage is the root cause that damages the reputation of the company. Employees of the company leak the information and put assets at risk. This paper described how emails are moving through intermediate computers before reaching their final destination, where confidential data is located and then apply the right set of controls to that data. This paper explained threats to an email and why there is a necessity to protect emails? The big advantage to deploy Email Protection System on Gateway is to protect company's sensitive information from the hands of competitors.

Our EPS maintains the three principles Confidentiality, Integrity and Availability to protect emails from malicious intent which violates the company's security. This Proposed System has the ability to block or quarantine inbound or outbound emails. We can't imagine a company that does not use an email for communication and equally it is difficult to think about a company that would not get any benefit from an email security gateway. Future efforts can be made in implementing the procedure that can tackle the real world scenario.

8. References

[1] A. Shabtai, Y. Elovici and L. Rokach, "A Survey of Data Leakage Detection and Prevention Solutions," Springer Briefs in Computer Science, Springer, 2012.

[2] J. Kumar and A. K. Singh, "Dynamic resource scaling in cloud using neural network and black hole algorithm," *2016 Fifth International Conference on Eco-friendly Computing and Communication Systems (ICECCS)*, Bhopal, 2016, pp. 63-67.

[3] H. Taneja, Kapil and A. K. Singh, "Preserving Privacy of Patients based on Re-identification Risk," *Fourth International Conference on Eco-friendly Computing and Communication Systems(ICECCS)*, 2015, pp. 448-454.

[4] I. Gupta, and A. K. Singh, "Privacy and Security Architecture for Cloud Data," technical report, NIT Kurukshetra, India, 2016.

[5] S. Chhabra and A. K. Singh, "Dynamic data leakage detection model based approach for MapReduce computational security in cloud," *2016 Fifth International Conference on Eco-friendly Computing and Communication Systems (ICECCS)*, Bhopal, 2016, pp. 13-19.

[6] I. Gupta, and A. K. Singh, "Data Leakage Detection in Cloud Environment," technical report, NIT Kurukshetra, India, 2017.

[7] S. Gill, G. Rupnar, V. Ramteke, D. Patil and V. M. Wadhai, "Email Security Protocol," *International Journal of Computer Trends and Technology (IJCTT)*, Mar-Apr 2011.

[8] P. Zilberman, S. Dolev, G. Katz, Y. Elovici and A. Shabtai, "Analyzing group communication for preventing data leakage via email," *Proceedings of 2011 IEEE International Conference on Intelligence and Security Informatics*, Beijing, 2011, pp. 37-41.

[9] R. Tahboub and Y. Saleh, "Data Leakage/Loss Prevention Systems (DLP)," *2014 World Congress on Computer Applications and Information Systems (WCCAIS)*, Hammamet, 2014, pp. 1-6.

[10] S. Peneti and B. P. Rani, "Data leakage prevention system with time stamp," *2016 International Conference on Information Communication and Embedded Systems (ICICES)*, Chennai, 2016, pp. 1-4.

[11] M. A. Faysel and S. S. Haque, "Towards Cyber Defense: Research in Intrusion Detection and Intrusion Prevention Systems," *International Journal of Computer Science and Network Security (IJCSNS)*, Vol.10, No.7, July 2010, pp. 316-325.

Creating adaptive web recommendation system based on user behavior

Bogdan Walek

Department of Informatics and Computers
30. dubna 22, Ostrava, 701 03
University of Ostrava

Email:bogdan.walek@osu.cz

Abstract. The paper proposes adaptive web recommendation system based on user behavior. The proposed system uses expert system to evaluating and recommending suitable items of content. Relevant items are subsequently evaluated and filtered based on history of visited items and user´s preferred categories of items. Main parts of the proposed system are presented and described. The proposed recommendation system is verified on specific example.

1. Introduction

Nowadays, there are a lot of web pages, web applications and web systems which contains various types of content (text, images, videos, etc.). Typically, web pages provide the same content for all users (online news portals, web portals, web magazines). Some of them adapt the content based on user behavior and user preferences (for instance Youtube, Booking.com, Facebook, etc.).

Adaptation of the content based on user behavior is a good benefit for user, because specific web page recommends suitable items of content or content categories, which are preferred by user.
Currently, there are few approaches for adaptive web pages, which can adapt the content.

The first approach is a rule-based system for adapt the web page content based on decision-making rules. The approach is more described [1], [2]. Another approach is a content-filtering system which filters the content of web page based on history of visited items [3]. Another approach is a collaborative filtering system which presumes that items preferred by user in past will be preferred also in future [4], [5], [6].

Another approach performs a prediction of the subsequent appropriate content items [7].
Current opportunities and challenges for adaptive web services are described in [8].
In this paper, we focus on web pages which contains products, typically e-shops.

2. Proposed approach

We propose an adaptive web recommendation system based on user behavior. The proposed system uses expert system to evaluating and recommending suitable items of content.

The proposed system recommends suitable items in two main situations, during first visit of user on web page and during the user´s activity at the web page (staying on the web page).

In the first situation, the system uses expert system to evaluate items relevancy based on item properties (importance of item, frequency of visiting the item by other users, frequency of selling the item by users). Evaluated items are then ordered by their relevancy and shown to users. This output decreases the effect called "cold start". Cold start is the main problem of current adaptive web systems and represents situation of no input data to these systems after first visit the web page by user [8]. Adaptive systems has no information about user behavior (history of visited items of content, time spent on specific web page, etc.), so it is difficult to recommend suitable items and adapt the content.

In the second situation, the system stores information about user behavior (history of visited items, preferred items, preferred categories of items, etc.) and adapts the content based on these data and items relevancy evaluated by expert system.

Main parts of the proposed approach will be described in the following sections.

2.1. Loading information about user

In the first step, information about user are loaded and prepared for subsequent steps. There are a lot of information about user which can be recognized and loaded during visit on the web page. Here are examples:

- Gender (male, female)
- Age
- Operating system type
- Browser type
- Device type (mobile phone, tablet, phablet, notebook, PC, etc.)
- History of visited pages on website

2.2. Creating the expert system

In the next step, an expert system is created. The knowledge base of expert system consists of IF-THEN rules. The main role of the expert system is to evaluate and recommend relevant content items to the user. After first visit of the web page, products will be evaluated and ordered by their relevancy. This step is important to decrease the effect of "cold start".

IF-THEN rules of expert systems consists of three input linguistic variables and one output linguistic variable.

Input linguistic variables are:

- IMP - importance of item in a category, administrator of we page decides which items are more (favorite items, bestsellers) or less important, possible values are:
    - low
    - medium
    - high
- VISIT - level of visiting the item – represents how many people visit the item, possible values are:
    - low
    - medium
    - high
- SALE - sales level of the item – represents frequency of selling the item, possible values are:
    - low
    - medium
    - high

Output linguistic variable is:

- RELEVANCY – level of item relevancy, possible values are:
    - very low
    - low
    - medium
    - high
    - very high

Examples of IF-THEN rules of knowledge base are shown below:

```
IF (IMP IS LOW) AND
(VISIT IS LOW) AND
(SALE IS LOW) THEN
(RELEVANCY IS VERY LOW)

IF (IMP IS MEDIUM) AND
(VISIT IS LOW) AND
(SALE IS HIGH) THEN
(RELEVANCY IS MEDIUM)

IF (IMP IS HIGH) AND
(VISIT IS MEDIUM) AND
(SALE IS MEDIUM) THEN
(RELEVANCY IS HIGH)
```

The expert system is created in LFLC (Linguistic Fuzzy Logic Controller) tool [9]. This tool enables to create a knowledge base of an expert system and also performs the process of the decision making together with the specific output, which is possible to use in the proposed system.

2.3. Evaluating items relevancy

In this step, the items of content are evaluated by expert system. The relevancy of each item is evaluated based on properties of items and knowledge base of the expert system.
Next, the items are ordered by their relevancy and recommended to the user. There are many ways how to show these items to the user. One of them is to create content block called "Relevant items" and put ordered items to this block. Evaluated items are displayed at the first and every subsequent visit of the web page.

Process of evaluating items relevancy by expert system is shown in Fig. 1.

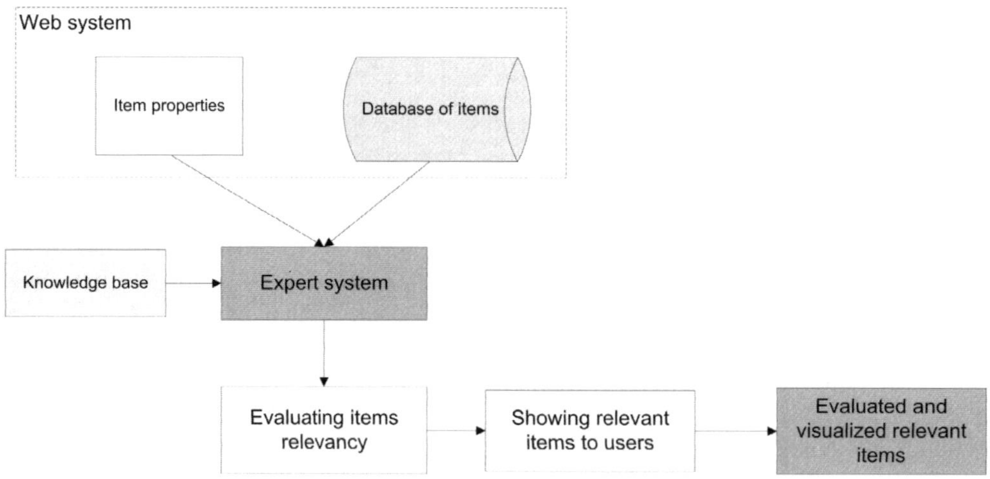

Figure 1. Process of evaluating items relevancy by expert system

IOP Conf. Series: Journal of Physics: Conf. Series **933** (2017) 012014 doi:10.1088/1742-6596/933/1/012014

2.4. Adapting the content and recommending the most relevant items

In this step, the content of the web page is adapted based on other user activity on the web. Based on visited items the system recommends the most relevant items based on these rules:

- Relevant items (items with highest relevancy evaluated by expert system) with the same categories as categories of items visited by user
- Relevant items with similar price as price of items visited by user

Based on these rules the system automatically selects the most relevant items and show them to the user ordered by their relevancy evaluated by expert system.

Process of recommendation of the most relevant items is visually shown in Fig. 2.

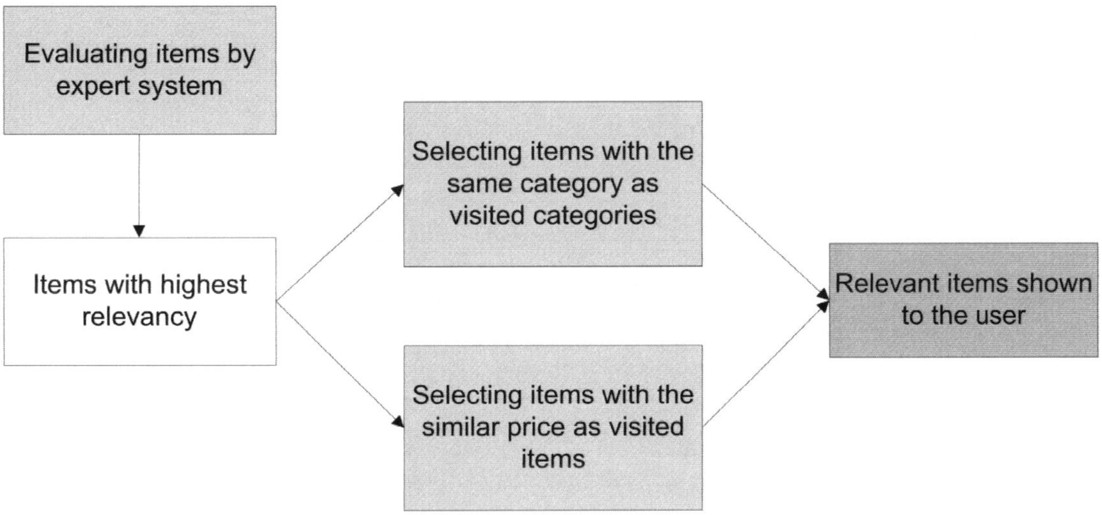

Figure 2. Process of recommendation of the most relevant items

3. Verification

For verification the selected products of real e-shop were chosen.

Information loaded by user are:

- Age
- History of visited pages on website

List of items for verification is shown in Table 1:

Num	Product name	Price (USD)	Category	Preferred age
1	The Miracle of Dunkirk	9.99	History	adults
2	Munich 1919: Diary of revolution	22.92	History	adults
3	Killing England	20.98	History	adults
4	The color of law	19.00	History	adults
5	The art of war	3.99	History	adults
6	A man called Ove	9.60	Literature & Fiction	adults
7	The secret she keeps	20.95	Literature & Fiction	adults
8	A gentelman in Moscow	16.32	Literature & Fiction	adults
9	The land of stories	13.11	Children´s books	children
10	The wonderful Things you	11.13	Children´s books	children

IOP Conf. Series: Journal of Physics: Conf. Series **933** (2017) 012014 doi:10.1088/1742-6596/933/1/012014

	Will Be			
11	A wrinkle in time	5.99	Children´s books	children
12	Women in science	11.55	Teen and young adults	teens
13	Michael Vey 7	13.98	Teen and young adults	teens
14	After the game	10.99	Teen and young adults	teens

Properties of products for evaluating by expert system with evaluated relevancy are shown in Table 2:

Num	IMP	VISIT	SALE	RELEVANCY
3	high	high	high	very high
6	high	high	medium	very high
10	high	medium	high	very high
11	high	medium	medium	high
4	high	medium	medium	high
14	high	low	high	high
8	medium	medium	medium	medium
13	medium	low	medium	medium
1	medium	medium	low	medium
2	low	medium	medium	low
7	low	medium	medium	low
12	low	low	high	low
5	low	low	medium	very low
9	low	low	low	very low

Items with evaluated relevancy are ordered and shown to the user during first visit of the web page. During browsing the e-shop these information were stored:

- User is adult.
- User visits the books with category History and Literature & Fiction.
- User visits the books with average price about 20 USD.

The system adapts the content and recommends these items ordered by their relevancy:

Num	Product name	Price (USD)	Category	Preferred age
3	Killing England	20.98	History	adults
4	The color of law	19.00	History	adults
8	A gentelman in Moscow	16.32	Literature & Fiction	adults
7	The secret she keeps	20.95	Literature & Fiction	adults
2	Munich 1919: Diary of revolution	22.92	History	adults

As we can see, the list of recommended items contains items of category History and Literature & Fiction, with similarity of price about 20 USD. Items are ordered by their relevancy.

4. Conclusion

In this paper we proposed the adaptive web recommendation system based on user behavior. The proposed system is connected to the expert system, which evalautes relevancy of content items. The

proposed system recommends relevant items of content based on evaluation by expert system and user behavior (history of visited items, preferred category).
Future work will focus on verification the proposed system on other examples and web pages. It will be also useful to generalize the approach and verify it on various types of web pages.

5. References

[1] Adomavicius G., Tuzhilin A. Expert-driven validation of rule-based user models in personalization applications. Data Mining and Knowledge Discovery, Vol. 5, Issue 1/2, 2001, pp. 33-58.
[2] Mobasher B. Data mining for web personalization. In The adaptive web. Springer Berlin Heidelberg, 2007, pp. 90-135.
[3] Van Meteren R., Van Someren M. Using content-based filtering for recommendation. In Proceedings of the Machine Learning in the New Information Age: MLnet/ECML2000 Workshop, 2000, pp. 47-56.
[4] Sarwar B., Karypis G., Konstan J., Riedl J. Item-based collaborative filtering recommendation algorithms. In Proceedings of the 10th international conference on World Wide Web, 2001, pp. 285-295.
[5] Perkowitz M., Etzioni O. Towards adaptive web sites: Conceptual framework and case study. Artificial intelligence, Vol. 118, Issue 1-2, 2000, pp. 245-275.
[6] Brusilovsky P., Peylo C. Adaptive and intelligent web-based educational systems. International Journal of Artificial Intelligence in Education, Vol. 13, 2003, pp. 159-172.
[7] Perkowitz M., Etzioni O. Adaptive web sites: an AI challenge. In IJCAI Vol. 1, 1997, pp. 16-23.
[8] Marquezan C. C., Metzger A., Pohl K., Engen V., Boniface M., Phillips S. C., Zlatev Z. Adaptive future internet applications: Opportunities and challenges for adaptive web services technology. In Application Development and Design: Concepts, Methodologies, Tools, and Applications, 2018, pp. 1568-1589.
[9] Habiballa H., Novák V., Dvořák A., Pavliska V. Using software package LFLC 2000. 2nd International Conference Aplimat 2003, Bratislava, 2003, pp. 355-358.

Acknowledgments

This work was supported by the project “LQ1602 IT4Innovations excellence in science” and during the completion of a Student Grant SGS08/UVAFM/2017 with student participation, supported by the Czech Ministry of Education, Youth and Sports.

IOP Conf. Series: Journal of Physics: Conf. Series **933** (2017) 012015 doi:10.1088/1742-6596/933/1/012015

A Peculiar Sentiment Analysis Advancement in Big Data

Manisha Valera[1], Yash Patel[2]

[1] Assistant Professor, Department of Computer Engineering, Indus University, Ahmedabad, Gujarat, India
[2] Student, Department of Computer Engineering, Indus University, Ahmedabad, Gujarat, India

Abstract. Every company wants to discover what its customers feel about it. But sentiment analysis can get coarser and turn inward to improve employee as well as customers' satisfaction. A term called sentiment analysis, or the mathematical taxonomy of statements' negative or positive connotations, gives companies potent ways to analyse cumulative language data across all sorts of communications. There's real value in measuring sentiment inside and outside your company. Our big data experts analyse the core grounds of the problems raised while handling giant datasets and find solutions to efficiently manage massive datasets with ease. We built a sentiment analysis solution to measure positive and negative sentiments with top rating products for Amazon.com's Electronics products as against competitor products.

1. Introduction

Sentiment Analysis involves recognizing the evaluative nature of a part of passage or text. For example, a product review can express a positive, negative, or unbiased sentiment. Automatically discovering sentiment articulated in text has a number of applications, including tracking sentiment towards products, movies, diseases etc., improving customer relation models, detecting pleasure and well-being. Over the past decade, there has been an extensive escalation in the use of micro blogging services such as Twitter and access to mobile phones world-wide. Thus, there is incredible curiosity in sentiment analysis of short informal texts, such as tweets and SMS messages, across a variety of domains. Now a days Internet is ruling over each and every aspect of the world, either selling or buying, ticket booking, ecommerce, education, medical field. Today, if a customer wants to buy a single product, He or she goes to ecommerce sites or application on mobile and check the Product's rating and review overall the world and determine which Product is bestselling in the compare of other brands. And then determine which one to buy. So, the review of the Product plays a vital role for getting sold the products. In our proposed algorithm we have taken Amazon data set of the Electronics Products and on that we have applied an algorithm to find out top rating products and on that we have applied sentiment analysis to find out the high positive sentiments from the customer reviews. So, the consequence of our algorithm will be highly selling Products all over the world in given data set which can be used for business analysis.

With the dawn of Big Data Analytics and Machine learning, it has been possible to create intelligent systems, which can analyse texts and derive insight which could be advantageous for business. In Ecommerce platforms with visitors hitting up to 2 billion, there is a lot of raw data that

IOP Conf. Series: Journal of Physics: Conf. Series **933** (2017) 012015 doi:10.1088/1742-6596/933/1/012015

can be analysed to understand the end user behaviour. If the user clicks, reviews, tags, blog posts, ratings all are gathered and processed many form of tortuous indicators can be converted to key performance indicators of a business thereby helping businesses to find target areas to be beneficial from manager perception. The customers are the solitary intention of any business; measuring their contentment level through these reviews can help both the customers and managers. Visuals for long have been making a lifelong impact on individuals, thereby empowering them in coming up with quick decisions. Providing them with visualizations and recommendations on the foundation of what others are saying can be relatively fruitful.

2. Related Work

In [1], presented a novel approach named Feature-Grading used to make the proposal of products in the e-commerce business. The procedure of Feature-Grading is isolated into five key strides: (a) Extracting general list of capabilities of a gathering classification of wares; (b)Extracting modifier set and negative words set; (c)Acquiring particular list of capabilities and highlight evaluation set; (d)Acquiring particular component weight set; (e)Acquiring thing weight set. The author has recommended the top ranking items using by applying rank and grade to all the items with an acquired grading equation.

A Proposed method in [2] is used to calculate sentiment score of reviews given by the customers is implemented in python on Hadoop. It uses Hadoop Distributed File System (HDFS) to store data set and run on Map Reduce architecture for performing sentiment analysis. The method works in two phases: In the first phase, Mapper will parse the given input file and in the second phase, Reducer will compute the sentiments. The data set they used contains different types of clothing product reviews from Amazon.com. Which present a statically a number of reviews about the product, number of positive and negative words, number of stop words. Stop word dictionary is used to recognize and remove stop words from the reviewed product. Increase/decrease the polarity strength when a word is preceded by a negation. Polarity Calculator uses positive, negative and stop-word dictionary to extract the sentiment-oriented words from each sentence. After calculating polarity the output will be stored in HDFS. Accuracy is performed by using the review scores of a product given by customers. The error rate for positive and negative reviews are comparatively little.

Shoushan [3] has performed a sentiment classification with polarity shifting phenomenon. They have extracted first the polarity shifted sentences using some detection rules. The same rules they have applied on testing data. And counting based classifier was designed based on polarity shifted words.

The author has proposed a Linguistic approach in [4], in that they have used emoticon symbols for separating the positive and negative comments in facebook. Thus, they have done Sentiment Lexicon analysis using sentiment words and classified positive and negative comments by prepared dictionaries.

The tweets were analysed in [5] to classify data and sentiments from Twitter more accurately. The information was extracted based on knowledge extraction. This is further enhanced using sphere specific seed based enhancement technique. The proposed method provides the extraction of keywords, entities, synonyms, and parts of speech from tweets which are then used for tweets classification and sentiment analysis. The proposed framework is tried on a gathering of 40,000 tweets. By applying the Knowledge Enhancer and Synonym Binder module on the extracted information the authors have achieved increase in information gain in a range of 0.1% to 55%.

In [6], it is intended to analyse social media big data to identify the widespread of certain keywords. A Java-Hadoop application is implemented to analyse data obtained from Twitter social network. The purpose was to identify number of people (Tweets) who mentioned specific medical keywords (e.g. Cancer) classified by location. The application is developed using Java Eclipse on CentOS Linux operating System and runs on Oracle Virtual Machine. It is aimed to help in decision making according to the number of people tweeting about cancer or any related word (like tumor) and analyse them according to their cities all over the world. The results are used to create a GIS layer to spatially boost the visualization of the obtained consequences.

The sentiment analysis system [7] that detects the sentiment of short informal textual messages such as tweets and SMS (message-level task) and the sentiment of a word or a phrase within a message. The system is based on a supervised statistical text classification approach leveraging a variety of surface form, semantic, and sentiment features. The sentiment features are mainly derived from tweet-specific sentiment lexicons. These lexicons are generated from tweets automatically with sentiment-word hash tags and from tweets with emoticons. To efficiently capture the sentiment of words in negated contexts, a separate sentiment lexicon is generated for negated words.

3. Proposed Work

In this section, a method to calculate the rating and based on that sentiment score of reviews given by the customers is proposed and implemented in python. This method works in two phases: In first phase, it will parse the given input file and in second phase, it will discover the average of each Product's rating and based on that it will find out the highest rating as outcome and calculate the sentiment score of reviews given by the Customer. The data set we used contains different types of Electronic products' rating and reviews from Amazon.com [8]. Some statistics about the data set are given in Table III. For instance, number of Products, number of Products with 5.0 rating, number of positive and negative sentiments, We use a positive and negative word dictionary to identify positive and negative words, Accuracy and Error rate [9]. Stop word dictionary is used to identify and remove stop words from the reviewed product [10].

4. Implementation and Result Analysis

The proposed approach is implemented on Amazon product reviews dataset [11]. We have implemented in python 3.5, the dataset which we have used is in JSON file format.


```
{
  "reviewerID": "A2SUAM1J3GNN3B",
  "asin": "0000013714",
  "reviewerName": "J. McDonald",
  "helpful": [2, 3],
  "reviewText": "I bought this for my husband who plays the
piano.  He is having a wonderful time playing these old
hymns.  The music  is at times hard to read because we think
the book was published for singing from more than playing
from.  Great purchase though!",
  "overall": 5.0,
  "summary": "Heavenly Highway Hymns",
  "unixReviewTime": 1252800000,
  "reviewTime": "09 13, 2009"
}
```


where

- reviewerID - ID of the reviewer, e.g. A2SUAM1J3GNN3B
- asin - ID of the product, e.g. 0000013714
- reviewerName - name of the reviewer
- helpful - helpfulness rating of the review, e.g. 2/3
- reviewText - text of the review
- overall - rating of the product
- summary - summary of the review
- unixReviewTime - time of the review (unix time)
- reviewTime - time of the review (raw)

Figure 1. Sample Review

We have converted JSON file format to CSV file format by removing unwanted columns and fetched only required columns like reviewer ID, Product ID, reviewer Text, Overall rating using Panda framework for increasing loading and reading speed. So, the file size would be reduced by 21.8%.

IOP Conf. Series: Journal of Physics: Conf. Series **933** (2017) 012015 doi:10.1088/1742-6596/933/1/012015

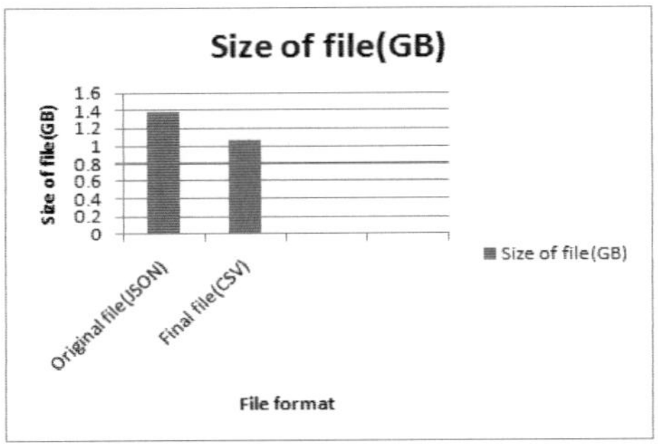

Figure 2. File size is reduced from JSON to CSV format

After converting to CSV format,

- Total Number of Reviews available: 1626187
- Total Number of Products available: 63000

```
C:\Users\Dell\AppData\Local\Programs\Python\Python35\python.exe E:/test/testpy.py
0528881469 4
0594451647 3
0594481813 6
0972683275 217
1400501466 41
1400501520 18
1400501776 18
1400532620 39
1400532655 107
140053271X 69
1400532736 11
1400599997 39
1400698987 26
1400699169 10
1615527613 3
3744295508 17
B00KSLCU72 11
B00KVNY2KA 8
B00KWHMR6G 12
B00KXAFYZS 4
B00KYMCJF8 14
B00L21HC7A 17
B00L2442H0 15
B00L26YDA4 18
B00L3YHF6O 19
Number of Products 63000
Number of Reviews 1626187
```

Figure 3. Product wise Number of Reviews

We read a CSV file's Product ID, Overall rating into pandas data frame.

- Product wise group the data and find the average Overall rating of individual Product.
- Fetch the List of Product ID with Overall rating is 5.0

After finding the Overall rating=5.0, we got total 2191 Product IDs. We have fetched reviewer Text for the products whose Overall rating=5.0.

- Analyze reviewer Text Product ID wise.
- Remove Stop Words.

IOP Conf. Series: Journal of Physics: Conf. Series **933** (2017) 012015 doi:10.1088/1742-6596/933/1/012015

Stop Words like which, after, has, that, become, in, before. are those words which are mostly used in a sentence and not of use in the sentiment analysis, so exclusion of these type of words will not affect the experimental results.

- Divide the reviewer Text into Positive and

  Negative Sentiments accordingly.

Usually negation will change the polarity of positive and negative words. For instance, "this product is not good" – in this sentence good is a positive word but if it is used with not it becomes negative[11]. So, we have given Weight age to each sentiment.

Table 1. Type of Sentiments

Positive Sentiments	Negative Sentiments
'good',' nice', 'excellent' ,'superb', 'awesome' ,'expensive', 'fine' ,'outstanding', 'magnificent' ,'fantastic', 'brilliant' ,'robust' ,'not bad'	'bad', 'not good', 'worst', 'cheap', 'inferior'

Table 2. Weightage for Sentiments

Sentiments	Weightage
nice, good, fine, robust	+1
excellent	+2
superb, fantastic, outstanding, magnificent, brilliant	+1.5
not bad	+0.9
bad, cheap, inferior	-1
not good	-0.9
worst	-2

- Sort the List of Product ID according to Weightage of sentiments.

  Using sorting we got the Products with Highest Positive Reviews which have Overall rating=5.

```
C:\Users\Dell\AppData\Local\Programs\Python\Python35\python.exe E:/amazon/readcsv.py
{'B005HQ4YSG': 5.0, 'B000S34B9U': 5.0, 'B005NEZ3XC': 5.0, 'B00AHWXS32': 5.0, 'B003WRG5DO': 5.0,
'B000VZ16AC': 5.0, 'B00ADXV1UC': 5.0, 'B003BYRLF4': 5.0, 'B001RI62XI': 5.0, 'B000E8VQ74': 5.0,
'B008XEYT6G': 5.0, 'B00140IDHW': 5.0, 'B0049VONBS': 5.0, 'B001MQ3CCO': 5.0, 'B004MM3LD0': 5.0,
'B008LSYCAC': 5.0, 'B002TJY5S8': 5.0, 'B0019EIBLQ': 5.0, 'B009I2S440': 5.0, 'B00066HJBG': 5.0,
'B000OU7EQE': 5.0, 'B00E4MQO8C': 5.0, 'B00006HXZV': 5.0, 'B0007LHVEU': 5.0, 'B0001Y7Z4Y': 5.0,
'B0009SZHR4': 5.0, 'B002ROE42M': 5.0, 'B005DSMPBQ': 5.0, 'B0002E3WIQ': 5.0, 'B0015FF92I': 5.0,
'B003DM0408': 5.0, 'B001C484LA': 5.0, 'B0042TJ7VI': 5.0, 'B004SQWY4W': 5.0, 'B000NMO90C': 5.0,
'B001F0PN3I': 5.0, 'B008S94NTE': 5.0, 'B0002J1LDE': 5.0, 'B00H7HH6QU': 5.0, 'B00008SCFU': 5.0,
'B000MXBR8O': 5.0, 'B000V0BXLE': 5.0, 'B002PDJMEK': 5.0, 'B00D8K4G54': 5.0, 'B0007TJDAW': 5.0,
'B0030M23RS': 5.0, 'B000BKV26A': 5.0, 'B0092CU7PK': 5.0, 'B00AHWXQDO': 5.0, 'B005K926MQ': 5.0,
'B000G6XY42': 5.0, 'B0077G7QDE': 5.0, 'B001QF5VCU': 5.0, 'B000SXWGAQ': 5.0, 'B002BSH2KA': 5.0,
'B00AXUWRUI': 5.0, 'B00003G1RC': 5.0, 'B00030DEOG': 5.0, 'B0057J86PK': 5.0, 'B000DZH712': 5.0,
'B00027BTAQ': 5.0, 'B00GMW49H0': 5.0, 'B00BVXHHLA': 5.0, 'B00023XDYA': 5.0, 'B004MX1ME4': 5.0,
'B0093E69HM': 5.0, 'B004NE448S': 5.0, 'B00443V7R4': 5.0, 'B000WUF67K': 5.0, 'B005D2C4KE': 5.0,
'B004FLE5W4': 5.0, 'B0011NPHWQ': 5.0, 'B00FN5UYO4': 5.0, 'B007KTQDRS': 5.0, 'B00382F12Y': 5.0,
'B000YJ0STU': 5.0, 'B00BAZPFMM': 5.0, 'B005EDDCW6': 5.0, 'B00B770TTC': 5.0, 'B002LVUISK': 5.0,
```

Figure 4. Product IDs with Overall Rating is 5.0

```
Number of products with high rating(5 stars): 2191
[('B00B4YE5BG', 31.75), ('B007BYLLNI', 31.65), ('B00L3YHF6O', 28.4), ('B000SDPMEI', 26.15),
('B00ELAIHDM', 24.0), ('B00HNJWV1W', 22.1), ('B002I9OBPG', 21.65), ('B000VRV6LY', 21.25),
('B002SQKVDO', 20.4), ('B0011NPHWQ', 20.0), ('B007JCNOHS', 20.0), ('B000INZRZW', 20.0),
('B0035FZ12O', 19.75), ('B00GNX8HSU', 19.4), ('B00K4VQZCM', 19.0), ('B0046EC1OE', 18.4),
('B002JI9FHU', 17.9), ('B002USVKFY', 17.65), ('B002IK8H0A', 17.65), ('B00318CGA0', 17.05),
('B003XDU47K', 16.95), ('B003E1WIPW', 16.9), ('B000F8YEFO', 16.65), ('B001TOD3KA', 16.3),
('B005JWIZC4', 16.1), ('B001FOR1G6', 16.0), ('B001TKRLRU', 15.75), ('B000FIJA6W', 15.75),
('B000G835DA', 15.65), ('B0051OLAWW', 15.65), ('B000TMFYBO', 15.649999999999999), ('B00A203I0W'
15.3), ('B007YWM09Q', 15.149999999999999), ('B00061LSAY', 15.0), ('B00005TFRK', 15.0), ('B00005
 15.0), ('B00LGQ6HL8', 14.75), ('B000DZFTZS', 14.75), ('B00JXLGE6Q', 14.399999999999999),
('B000FDMAHS', 13.75), ('B007JY58FM', 13.65), ('B004A8PTAE', 13.65), ('B0026FCKA0', 13.65),
('B003U4VIWW', 13.65), ('B00DKAWM80', 13.65), ('B00AY6MGRU', 13.5), ('B0013545U2', 13.0),
('B00622STI0', 13.0), ('B003O5LLFG', 13.0), ('B002KOWDAE', 13.0), ('B0052GVNCQ', 13.0),
('B00GYFLO14', 12.95), ('B000G1NWYO', 12.75), ('B00HUGNN06', 12.75), ('B002NPCDP0', 12.65),
('B003VANRAQ', 12.65), ('B00BLO5GZS', 12.65), ('B0000E68U2', 12.55), ('B0007WKB0K', 12.3),
('B002HHLGD4', 12.05), ('B00DQXJM6Q', 12.0), ('B005K8CSZ2', 12.0), ('B0061U0SUK', 12.0),
('B00AE991VG', 12.0), ('B000UPA4R4', 12.0), ('B004MKNQHI', 11.95), ('B003B56LFY', 11.75),
('B000F7A85A', 11.75), ('B000CZ0RL0', 11.65), ('B0034XJ3TY', 11.65), ('B00CGY4ET6', 11.65),
('B004I43MK4', 11.5), ('B00005ABC9', 11.4), ('B003D8QQ5Y', 11.0), ('B000EZKAHY', 11.0),
('B007O0KR48', 11.0), ('B00JGGV15C', 11.0), ('B000PH8XI8', 11.0), ('B000GS073M', 11.0),
('B00K2L3G5I', 11.0), ('B00192KEY0', 10.65), ('B00HPPIOW8', 10.65), ('B0011YR8Q8', 10.65),
```

Figure 5. Sortest List of Product ID with Highest Review Weightage and Overall rating=5.0

Though Our Algorithm's outcome has majority of Positive values for the sentiments, there is some human error for the sentiments which are having top rating for the products. Therefore, some products' review shows negative value in our prediction.

Table 3. Results after Implementation.

Number of products	63000
Number of products with 5.0 rating	2191
Number of products with positive comments	2085
Number of products with negative comments	106
Accuracy (%)	95.16
Error (%)	4.84

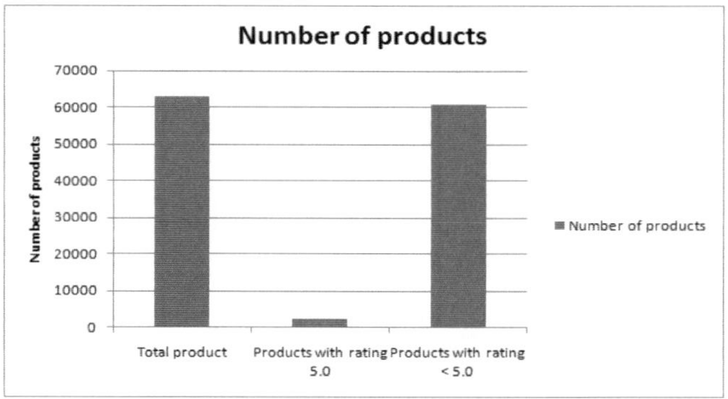

Figure 6. Number of Products with Overall rating=5.0

After Implementation of this algorithm, A test of the sentiment analysis has been conducted to quantify accuracy. There would be 95.16% accuracy in finding out which Product is best sold based on rating and review. But, due to human error in reviews we got error of 4.84%.

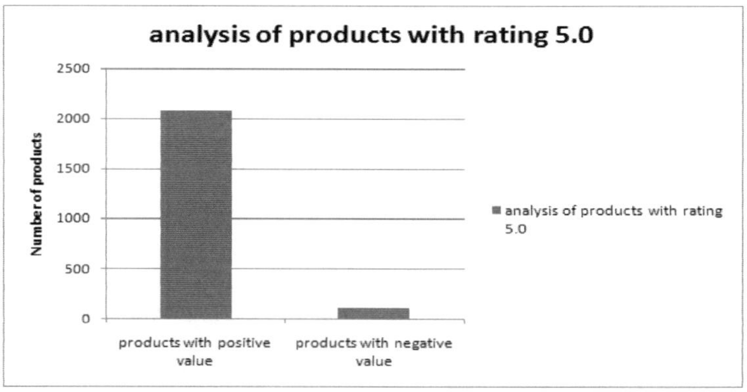

Figure 7. Number of Products with Overall rating=5.0 and Highest Positive Sentiments

5. Conclusion

Using Proposed Algorithm, We can find out the best selling product based on the rating as well as by sentiment analysis on reviews given by user, since first of all we have extracted the products with top rating, that is 5 star rated and after that by applying sentiment analysis on top fetched product for finding the most best reviewed products of all top rated product and sorting the products according to their Weightage of sentiment types, we got 95.16% accuracy. That means we are 95.16% successful in understanding human sentiment regarding top products and hence predicted the best product of all product. For further advancement in finding best product we can use machine learning algorithm and Natural Language processing for detecting weightage for sentiment keywords.

6. References

[1] Luo Yi; Fan Miao; Zhou Xiaoxia, *The design and implementation of Feature-Grading recommendation system for e-commerce*, Information and Automation (ICIA), 2011 IEEE International Conference on, vol., no., pp.236, 241, 6-8 June 2011

[2] Piyush Gupta, Pardeep Kumar, and Girdhar Gopal. *Article: Sentiment Analysis on Hadoop with Hadoop Streaming*. International Journal of Computer Applications 121(11), 4-8, July 2015

[3] Shoushan Li; Zhongqing Wang; Lee, S.Y.M.; Chu-Ren Huang, *Sentiment Classification with Polarity shifting Detection*, Asian Language Processing (IALP), 2013 International Conference on , vol., no., pp.129,132, 17- 19 Aug. 2013

[4] Jamoussi, S.; Ameur, H., *Dynamic construction of dictionaries for sentiment classification*, Cloud and Green Computing (CGC), 2013 Third International Conference on , vol., no., pp.418,425, Sept. 30 2013-Oct. 2 2013

[5] Batool, R.; Khattak, A.M.; Maqbool, J.; Sungyoung Lee, *Precise tweet classification and sentiment analysis*, Computer and Information Science (ICIS), 2013 IEEE/ACIS 12th International Conference on , vol., no., pp.461,466, 16-20 June 2013

[6] Alenezi, Sultan, and Saleh Mesbah. , *Big Data Spatial Analytics in Social Networks using Hadoop*, International Journal of Computer Applications 128.14 (2015): 21-26.

[7] Svetlana Kiritchenko, Xiaodan Zhu, Saif M. Mohammad, *Sentiment Analysis of Short Informal Texts*, Journal of Artificial Intelligence Research 50 (2014), 723–762

[8] jmcauley.ucsd.edu/data/Amazon/

[9] Minqing Hu and Bing Liu., *Mining and Summarizing Customer Reviews*, Proceedings of the ACM SIGKDD International Conference on Knowledge Discovery and Data Mining (KDD-2004), Aug 22-25, 2004, Seattle, Washington, USA.

[10] "stop-words," (Last visited in June 2015) [online]. Available: https://code.google.com/p/stop-words/

[11] Godbole, N., Srinivasaiah, M., & Skiena, S., *Large-Scale Sentiment Analysis for News and Blogs*. ICWSM 7 (2007): 21

IOP Conf. Series: Journal of Physics: Conf. Series **933** (2017) 012016 doi:10.1088/1742-6596/933/1/012016

Pithy Review on Routing Protocols in Wireless Sensor Networks and Least Routing Time Opportunistic Technique in WSN

Mohammed Salman Arafath[1], Khaleel Ur Rahman Khan[2], K.V.N Sunitha[3]

[1]Dept of CSE, Keshav Memorial Institute of Technology Hyderabad, Telangana, India
[2]Dept of CSE, ACE College of Engineering, Hyderabad, India
[3]Dept of CSE, BVRIT College of Engineering, Hyderabad, India

Abstract. Nowadays due to most of the telecommunication standard development organizations focusing on using device-to-device communication so that they can provide proximity-based services and add-on services on top of the available cellular infrastructure. An Oppnets and wireless sensor network play a prominent role here. Routing in these networks plays a significant role in fields such as traffic management, packet delivery etc. Routing is a prodigious research area with diverse unresolved issues. This paper firstly focuses on the importance of Opportunistic routing and its concept then focus is shifted to prime aspect i.e. on packet reception ratio which is one of the highest QoS Awareness parameters. This paper discusses the two important functions of routing in wireless sensor networks (WSN) namely route selection using least routing time algorithm (LRTA) and data forwarding using clustering technique. Finally, the simulation result reveals that LRTA performs relatively better than the existing system in terms of average packet reception ratio and connectivity.

1. Introduction

Wireless Sensor Network (WSN) is a billion dollar market. WSN is the backbone of emerging technologies like the internet of things IoT, vehicular networks, cyber-physical systems etc. Wireless sensor networks are developed to support a huge range of applications ranging from home automation to smart battlefields [1]. Wireless sensor networks consist of a huge number of tiny sensor nodes that are arranged in spatially distributed terrain. Sensor node can sense, store, transmit and receive data from environments [2] while designing of wireless sensor network faces the major issues are energy consumption, security, and routing.

This paper targets routing part i.e. opportunistic routing and other important metrics are not focused here. Routing in wireless sensor network is one of the paramount aspects. The nitty-gritty of wireless sensor network is broadcasting nature of the network. This fact is used by the opportunistic routing.

2. Opportunistic Routing

Routing in WSN using Opportunistic routing is a newfangled paradigm to increase the network lifetime, packet delivery ratio or packet reception ratio. Routing in wireless sensor network (WSN) is an onerous task as it involves many parameters. There are many traditional routing algorithms for wireless sensor networks however it has been proven that Opportunistic routing is better and more efficient than the traditional routing [3]. If routing solution is good for WSN then it results in increased

IOP Conf. Series: Journal of Physics: Conf. Series **933** (2017) 012016 doi:10.1088/1742-6596/933/1/012016

in network lifetime and decreased in energy consumption otherwise the results will be contrary. Therefore this field has attracted many researchers.

2.1. Concept of Opportunistic Routing (OR)
WSN utilizes the shared random access wireless channel and most of the routing protocols consider this wireless medium as a limitation due to temporal variation. This shared wireless medium which was an obstruction to most of the routing protocols is an opportunity in an Opportunistic Routing (OR). The essential concept of OR it to exploit the broadcast nature of wireless medium. When OR advertise a data packet then that transmission of one node is overheard by multiple neighbour nodes which later forms the forwarder set. The crucial task of OR is to select a packet forwarder node, this is achieved by OR not by pre-selecting next forwarder node well ahead of time but the next forwarder node is selected dynamically at the time of transmission. The forwarder node is selected from the forwarder set in such a way that it is closest to the destination. The ExOR Opportunistic Multi-Hop Routing for Wireless Network [4] is the pioneer in practically implementing the OR in wireless networks as it dynamically chooses the paths on a per transmission basis in a wireless network

2.2. Pithy Review of Routing Protocols in Wireless Sensor Networks and role of OR
Networks are meant so that packets can be routed. Here our focus is WSN but there is another category of the networks known as Opportunistic Networks or Opportunistic sensor networks (Oppnets). The differentiating criterion between these two is the intermittent connection. Since OR takes advantage of Opportunistic contacts in wireless shared medium all the routing protocols in Oppnets should have the ability to withstand with the network disconnection. We may say that routing protocols in Oppnet are Opportunistic in nature and they are in the category of OR. However, the routing protocols in WSN are not opportunistic in nature because WSN is well-connected network and the intermittent connection is an exception here. Incorporating the concept of opportunity in WSN shown to the world that it has lot of improvements in term of battery lifetime, packet delivery ratio, increasing network lifetime etc; Since WSN have shared wireless medium and flooding is used to broadcast messages to the nodes in WSN the point is that we need to design mechanisms and techniques to control this flooding and we term it as forwarding. So opportunistic routing is nothing but controlled flooding by various techniques and mechanisms. Table 1 below shows in chronological order the various algorithms used in OR are nothing but controlled flooding and a mix of GPS (it is also used to control flooding messages) so that messages are not sent in the wrong direction of destination to avoid energy consumption.

Table 1. Comparison of OR Algorithms based on forwarding technique with merits and demerits

Reference/ Year	**Forwarding Technique/ controlled flooding**	**Merits and Demerits / Remarks**
Reference[5] 2017	Based on Path Protected Hop by Hop Protocol. A node having no certificate avoided from the route and another path will be selected	Improved security, throughput, energy consumption, packet overhead and network reliability.
Reference[6] 2017	Based on ML probability in terms of successful delivery, dropped messages, hop counts and overhead	It outperforms PROPHET+ at cost of small buffer size and space.
Reference[7] 2016	Based on combining the concept of Opportunistic routing with rateless coding and Orthogonal Frequency Division Multiple Access	It outperforms the strategy of forwarding along unique path by up to 77%
Reference[8] 2015	Based on forwarder list and the nodes in the forwarder list prioritizing themselves based on optimal hop and residual energy distance	Better energy usage and increased network lifetime. Packet delivery rate is better than reference [10]
Reference[9] 2013	Based on the cost metric such as social profile, social connectivity and social interaction of node	Efficient routing and increased packet delivery.
Reference[10] 2012	Based on trust value(direct and indirect) and cost distance from destination	Reduced End to End packet delay and increased in security gains.
Reference[11] 2011	Based on forwarding set which in turn selected based on total expected cost TEC. TEC is calculated as min of	For each node expected cost is updated using Bellman-Ford algorithm.

	expected cost of node to be added or the prefix forwarding list	
Reference[12] 2011	The only destination node can opportunistically receive the packet by overhearing the transmission of nodes in the traditional network and sends destination ACK to other nodes on the path.	Reduces packet duplication rate Increases throughput.
Reference[13] 2010	Based on forwarder list in which nodes prioritize them based on a cost metric. Cost metric which is based on Optimal link transmission aspect, optimal path metric and node metrics.	Better throughput, improved end to end delay performances compared with traditional computer network algorithms.
Reference[14] 2009	Priority candidate for the next hop among the nodes depends on priority order and delivery ratio.	Good for analyzing the performance parameters
Reference[15] 2009	Always low priority nodes hear the advertisement of high priority nodes	No duplications
Reference[4] 2005	It identifies the shortest path to destination and ties are resolved using delivery ratio matrix	It outperforms the best-predetermined routes in number of transmissions
Reference[16] 2003	Based on Geographic routing. When sender node broadcast its own and destination location the listening nodes will prioritize themselves to act as a relay node.	Analysis of performance can be done in terms of a number of hops to reach the destination.
Reference[17] 1998	It uses geographic information to direct packets towards the destination.	It is reminiscent of flooding

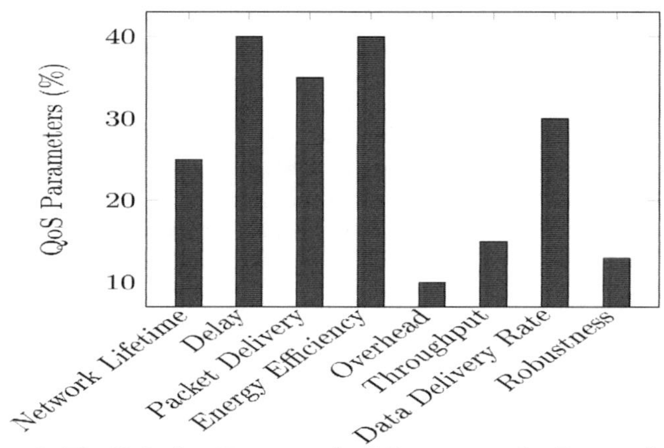

Figure 1. The Relative Focus on QoS Parameters by Researchers [18].

In this paper focus is only on routing taking packet delivery ratio as the important quality of service parameter according to the author of [18], as shown in figure 1 above the packet delivery is a second highest parameter and this paper is addressing the same metric as shown in the figure below which is reproduced from [18].

The data shown in figure 2 below represents the recent research and developments in the field of routing in WSN [18]. The routing protocols have been developed to face many challenges and one of them is packet reception and packet delivery, which is the prime focus of this paper. When compared with the year 2015 in 2016 the factor of developing routing solutions are fall by a factor of 5.

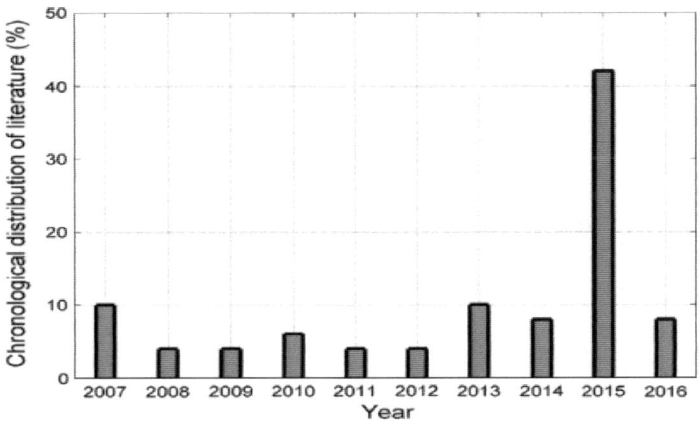

Figure 2. Development of Routing protocols for WSN over the time period [18].

3. Proposed Architecture

As shown in figure 3 below in order to provide efficient communication between source nodes to the destination node, clustering of the nodes will be performed and among the clusters, a cluster head is selected based on the clustering algorithm, an i.e. node which is nearest to the destination node. From the available best cluster heads the source node chooses to send the packet to cluster heads using intermediate nodes on which it trusts most. The cluster head, in turn, communicates with the destination node. For the sake of understanding in figure 3 different colors are used for selected neighboring node, cluster head, cluster member, the source node and the destination node. Similarly, the arrows in figure 3 show the available path and path selected for routing, request messages sent from cluster heads and response messages from sensor nodes. Clustering and Routing pseudo-code is given in the subsequent sections. Finally, the simulation result shows the number of data packets received is relatively better than the existing system [19].

3.1. Identifying Problems and Proposed Solutions

The author of the paper [20] proposes Establishing Stable and Reliable Routes algorithm i.e. establishing a stable and reliable route in multi-hop heterogeneous wireless sensor networks. It combines the payment and trust management with trust values and energy aware routing protocol. This routing protocol selects routing path based on trust value, which calculates trust value based on past behaviors of a node in the different time interval. This paper provides the two routing protocols Shortest Reliable Route (SRR), and Best Available Route (BAR) for reducing energy consumption in networks. However, the drawback is routing is not efficient because this algorithm calculates the trust value based on the behavior of node only. But the behavior of node may change into malicious. So it is not efficient for routing. In our proposed solution trust value is calculated based on packet reception ratio, it's efficient for routing.

The author in [21] presents the active trust routing protocol for secure routing in sensor networks. This protocol provides a quick detection of black hole attack in networks by using the trust values of nodes. Active trust algorithm significantly improves the data route access probability and ability to detect a black hole attack. However, path selection is difficult, due to source node selects only high trust value with the nearest node to the destination. The proposed solution is to select the path based on a best available path which is least routing time path.

3.2. Clustering Nodes

Clustering is the process of organizing the sensor nodes in groups. In this clustering algorithm sensor node forms the group based on least distance from the destination as shown in section 3.2.1.

Pseudo Cod e for Clustering Nodes. The following is the pseudo code for clustering nodes.

i. Begin
ii. Destination node gets messages from all the nodes
iii. Destination node computes minimum distance to all other best nodes
iv. Search for node closest to another node, if one of them is close to the border remove it.
v. Destination node announce the cluster head and advertise to all the nodes
vi. Cluster head is formed based on least distance with the destination and having higher energy
vii. Send data to cluster head and cluster head sends data to destination
viii. If cluster head has low energy or least distance choose which is nearest to it
ix. New cluster head will communicate with the destination node
x. End.

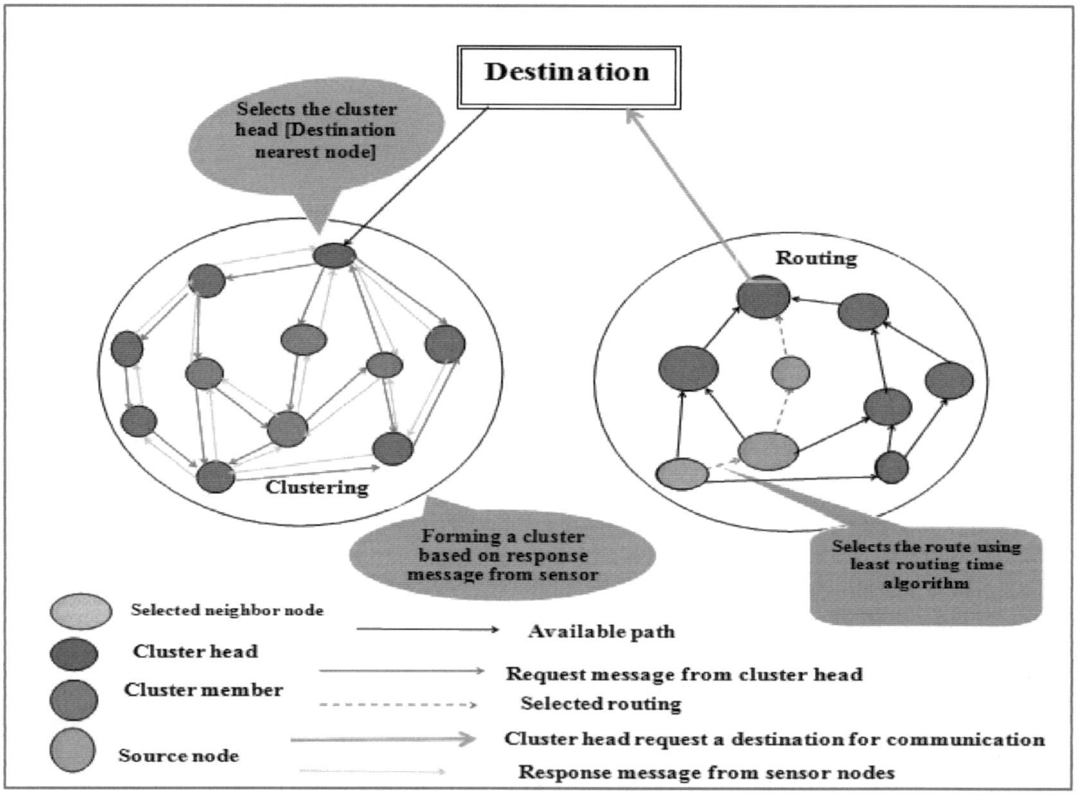

Figure 3. The Proposed Architecture

3.3. Routing

In routing forwarding, a packet to the intermediate node is one of the crucial decisions. To help source node to take the decision to trust the intermediate node this algorithm introduces the concept of trust value. All the intermediate nodes which can carry packet from source to cluster head may have different trust values or no trust value. The one with highest trust value will be selected for intermediate node. This trust is calculated based on packet reception ratio as shown in section 3.3.1.

Pseudo Code for Routing. The following is the pseudo code for routing.

i. Begin
ii. All nodes providing request
iii. All these requests are sent to cluster by the nodes

iv. Calculate Packet Reception ratio for all the nodes.
v. If Packet reception ratio is high then trust value is assigned
vi. Otherwise, no trust value is assigned.
vii. Sender nodes select the intermediate node based on high trust value
viii. Sender nodes sends packet to cluster head
ix. Cluster head verifies it
x. If packet is valid it moves packet to the destination
xi. Otherwise, drops the packet
xii. End

3.4. Packet Delivery Ratio

In this section packet delivery ratio on the proposed solution is described. Packet delivery ratio (PDR) is the ratio of total packets successfully sent to the source node to the total packet received by the destination node. The proposed solution results in reveals that the packet delivery ratio is better compared with the existing researches. The table 2 below depicts that when 50 nodes are taken then existing system gives 95% of packet delivery ratio whereas the proposed system gives 100% of packet delivery ratio. Similarly when 100,150 and 200 nodes are taken the respective packet delivery ratio in the existing system is 89%, 82%, and 79% respectively. However, the proposed system gives 100%,95% and 94% of packet delivery ratio for 100,150,200 nodes respectively. Figure 4 shows the aforementioned results in graphical form.

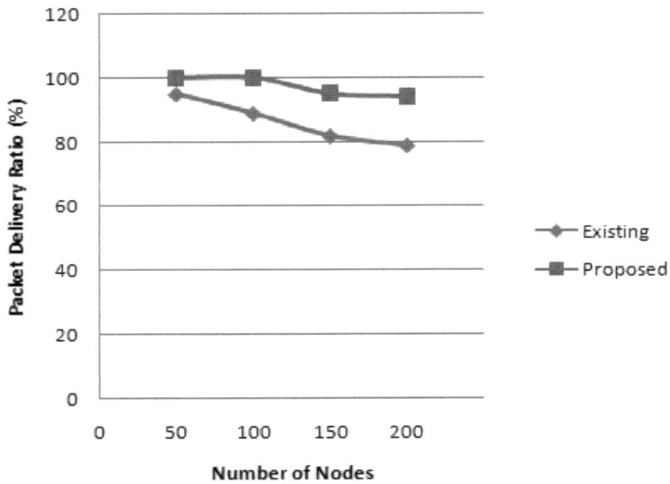

Figure 4. Comparison of Number of packets received with the existing system [19].

Table 2. Values of Number of Packets received

Number of Nodes	Existing[19]	Proposed
50	95	100
100	89	100
150	82	95
200	79	94

The proposed solutions of opportunistic routing in WSN are implemented in OMNeT++ simulation software. Omnet++ is open source. The standard IEEE 802.15.4 is designed for various supports in WSN one of them is routing also.

4. Conclusion, Future Work and Scope

In current work, we focus only on routing and packet delivery ratio. However, this paper is successful in getting the better results relatively with the existing system. Improvement in packet delivery ratio makes an attack on availability more difficult. It is possible to extend this work to incorporate security in OR (Opportunistic Routing) further the work can be extended by analyzing the replication problem and including some more data structure, it is possible to avoid replication attacks. Still, there is the scope of efficient energy management in WSN. The proposed algorithm is portable and it may be deployed in Opportunistic sensor networks to get the same results.

5. References

[1] Mohammed Salman Arafath, Khaleel Ur Rahman Khan, " Opportunistic Sensor Networks: A Survey on Privacy and Secure Routing", 2017 2nd IEEE International Conference on Anti-Cyber Crimes (ICACC), pp.41-46, 26-27 March 2017.

[2] Sukhwinder Sharma, Rakesh KumarBansal, Savina Bansal, " Issues And Challenges in Wireless Sensor Networks" IEEE International Conference, pp 1-5, 2013.

[3] Sanjit Biswas, Robert Morris, "Opportunistic Routing in Multi-Hop Wireless Networks" MIT Laboratory for Computer Science.

[4] S. Biswas and R. Morris, "ExOR: Opportunistic multi-hop routing for wireless networks," in Assoc.Comput. Mach. SIGCOMM Comput. Commun. Rev., 2005, vol. 35, no. 4, pp. 133–144.

[5] Arafath, M.S., and Khan, K.U.R. 'Incorporating privacy and security in military application based on opportunistic sensor network', Int. J. Internet Technology and Secured Transactions, (This paper has been selected for publishing the Complete Reference will be provided soon) http://www.inderscience.com/info/ingeneral/forthcoming.php?jcode=IJITST

[6] D. K. Sharma; S. K. Dhurandher; I. Woungang; R. K. Srivastava; A. Mohananey; J. J. P. C. Rodrigues, "A Machine Learning-Based Protocol for Efficient Routing in Opportunistic Networks," in IEEE Systems Journal, vol.PP, no.99, pp.1-7 doi: 10.1109/JSYST.2016.2630923.

[7] F. Hohmann and A. Klein, "Opportunistic Forwarding Using Rateless Codes in OFDMA Multihop Networks," 2016 IEEE 84th Vehicular Technology Conference (VTC-Fall), Montreal, QC, 2016, pp. 1-5.

[8] Juan Luo, Member, IEEE, Jinyu Hu, Di Wu, Member, IEEE, and Renfa Li, Senior Member, IEEE "Opportunistic Routing Algorithm for Relay Node Selection in Wireless Sensor Networks" IEEE Transactions on Industrial Informatics, VOL. 11, NO. 1, February 2015.

[9] Gary K. W. Wong, Xiaohua Jia," A Novel Socially-Aware Opportunistic Routing Algorithm in Mobile Social Networks" International Conference on Computing, Networking and Communications, Wireless Ad Hoc and Sensor Networks Symposium 2013.

[10] Jie Zhingzhong, Huan chuanhe, Xu Liya, Wang Bo, Chen Xi, Fan Xiying."A Trusted Opportunistic Routing Algorithm for Vanet" Third International Conference on Networking and Distributed Computing 2012.

[11] Xiaohua Xu, Student Member, IEEE, Xiang-Yang Li, Senior Member, IEEE, and Huadong Ma, Senior Member." Energy-Efficient Opportunistic Routing in Wireless Sensor Networks" IEEE Transaction on Parallel and Distributed Systems, VOL. 22, NO. 11, Nov 2011.

[12] Goo Yeon Lee and Zygmunt J. Haas,"Simple, Practical, and Effective Opportunistic Routing for Short-HaulMulti-Hop Wireless Networks", IEEE Transaction on Wireless Communication, VOL. 10, NO. 11, Nov 2011.

[13] Shih-Chun Lin and Kwang-Cheng Chen"Spectrum Aware Opportunistic Routing in Cognitive Radio Networks". IEEE Globecom 2010 proceedings.

[14] Angela Sara Cacciapuoti, Marcello Caleffi, and Luigi Paura "A theoretical model for opportunistic routing in ad hoc networks "Ultra Modern Telecommunications & Workshops, 2009. ICUMT '09.

[15] Y. Li, W. Chen and Z. L. Zhang, "Optimal forwarder list selection in opportunistic routing," 2009 IEEE 6th International Conference on Mobile Adhoc and Sensor Systems, Macau, 2009, pp. 670-675.doi: 10.1109/MOBHOC.2009.5336939.

[16] M. Zorzi and R.R. Rao, "Geographic Random Forwarding (GeRaF) for Ad Hoc and Sensor Networks: Energy and Latency Performance,"IEEE Trans. Mobile Computing, vol. 2, no. 4, Oct.-Dec. 2003.
[17] Young-Bae Ko and Vaidaya Nitin H. Location-Aided Routing (LAR) in mobile ad hoc networks. In Proc. ACM/IEEE MobiCom, pages 66-75, October 1998.
[18] A. Sarkar, T. S. Murugan, "Routing protocols for wireless sensor networks: What the literature says?", Alexandria Eng. J., vol. 55, pp. 3173-3183, Dec. 2016.
[19] L. S. Sindhuja and G. Padmavathi, "Replica Node Detection Using Enhanced Single Hop Detection with Clonal Selection Algorithm in Mobile Wireless Sensor Networks", HindawiResearch article, Journal of Computer Networks and Communications, pp. 1- 14, 2016.
[20] Mohamed M. E. A. Mahmoud, Xiaodong Lin, and Xuemin (Sherman) Shen, "Secure and Reliable Routing Protocol for Multihop Heterogeneous Wireless Sensor Networks" IEEE Transactions on Parallel Distributed Systems, pp 1-11, 2013.
[21] Yuxin Liu, Mianxiong Dong, Member, Kaoru Ota, Member, AnfengLiu,"Active Trust Secure Trustable Routing In Wireless Sensor Networks", IEEE Transactions on Information Forensics and Security, pp 1-14, 2016.

IOP Conf. Series: Journal of Physics: Conf. Series **933** (2017) 012017 doi:10.1088/1742-6596/933/1/012017

Design of Simple Landslide Monitoring System

Meng Qingjia, Cai Lingling

1305 QiyiiZhong Road, Jingxiu District, Baoding City, Hebei Province, China

menmen-00@163.com; 724831589@qq.com

Abstract. The simple landslide monitoring system is mainly designed for slope, collapse body and surface crack. In the harsh environment, the dynamic displacement data of the disaster body is transmitted to the terminal acquisition system in real time. The main body of the system adopt is PIC32MX795F512. This chip is to realize low power design, wakes the system up through the clock chip, and turns on the switching power supply at set time, which makes the wireless transmission module running during the interval to ensure the maximum battery consumption, so that the system can be stable long term work.

1. Introduction

In China, a series of disasters are caused by geological disasters, especially landslides. Natural weathering, water erosion, engineering construction, which making the slope of the force is damaged, landslide displacement is prone to occur. In recent years, there have been many technical means for remote landslide disaster monitoring in China, and the dynamic changes of landslide hazard are monitored by computer, remote sensing and other technical means in real time. From the two aspects hardware selection and software design, the landslide disaster monitoring technology continues to achieve integrated development, and gradually tends to high precision, improving the security and stability of the system.

The simple landslide monitoring system mainly adopts the parameter of landslide displacement to design, selects PIC32MX795F512L as the main chip to realize the low power design, and satisfies the long-term field operation. The system has been increased the self-test, error function and the control terminal to feedback the information resetting or restarting which achieves part of the intelligent. The system has increased the anti-theft warning. It is designed to avoid the human factors bring about by the failure of the equipment causing the lack of data. So it can protect the monitoring equipment and data integrity. At the same time, for the field of landslide disaster monitoring early warning project, it provides a more reliable technical services and support.

2. System composition and characteristics

The simple landslide monitoring system sends the displacement parameters of the landslide hazard and the operation of the instrument to the monitoring terminal. The system mainly carries out data collection, analysis and processing, electricity and anti-theft inspection. Through the 4G network, it transmits the information to the terminal. The receiving terminal feedbacks signal to achieve parameter reset, sleep, and inspection and so on. The system mainly implements the data sending, receiving and forwarding function. With the 4G module of Huawei, the on-site monitoring data is sent

IOP Conf. Series: Journal of Physics: Conf. Series **933** (2017) 012017 doi:10.1088/1742-6596/933/1/012017

to the PIC32MX795F512L through the protocol RS232, and the data is sent to the data monitoring terminal through the 4G network.

2.1. *System hardware*

2.1.1. *Data processing unit.*
Using the minimum power theory, the system uses the commonly used PIC32MX795F512L master chip. The chip has USB, CAN and Ethernet high-performance 32-bit flash memory, multiple power management modes and independent configurable watchdog timer. The chip has a low power RC oscillator and fault protection monitor. The chip supports RS-232. It has hardware decoding 6 UART, 5 external interrupt pins, many I / O pins which can be configured for missing open outputs etc... The application of the chip is to ensure that the simple landslide monitoring system is more reliable. Setting the real-time clock and calendar, the displacement of the landslide disaster body can be monitored. The monitoring data is larger than the initial value stored in the chip and is lower than the early warning value. The early warning value is the current value - initial storage value in the chip. The stored value in the original chip is discarded and replaced. Simultaneously, PIC32MX795F512L gives an order to start the 4G module. The 4G module sends alarm information to the control center. Using ultra-low power built-in amplifier and reference voltage AD7793 analog-to-digital conversion chip, the common mode signal interference can be effectively reduced. The signal error is also reduced. The system can get the displacement of digital signals and voltage signals more precisely. Once the data's transmission delivery failed three times, the monitoring data will be stored in the external TF memory card. When the monitoring terminal can't control and receive the information from the monitoring equipment, the staff will extract the data to the monitoring point to prevent the data from losing. (Figure 1)

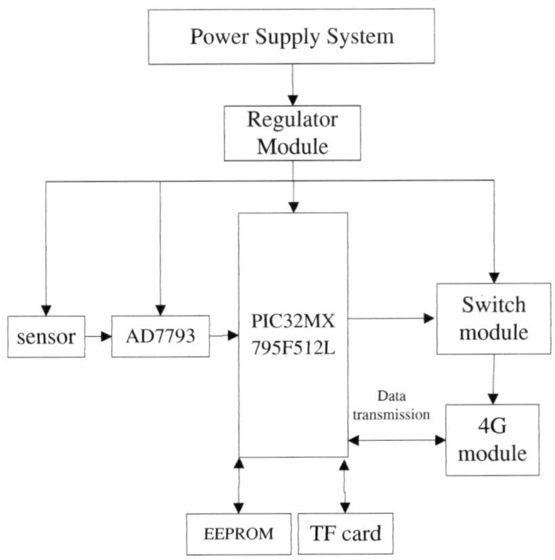

Figure 1 Hardware system

2.1.2. *Power supply system.*
In the landslide monitoring system, it is designed to use 12V solar charging controller to protect the 12V lead-acid batteries, to avoid over-voltage damage to the battery. The system uses 60W thin film polysilicon solar panels which are Easy to be carried and installed in the field. The solar panel's peak voltage is 17.2V. Its short circuit current is 3.85A;this kind of solar panels is compression which suitable ambient temperature is -40 ℃ ~ 85 ℃ . Because of the difference voltage of PIC32MX795F512L, AD7793, 4G modules and other power requirements, the power supplies are

various.Using LM2596S, the voltage converts from 12V to 5Vfor the displacement sensor. LM117-3.3 converts the voltage from12V to 3.3V for other hardware

2.1.3. 4G module.

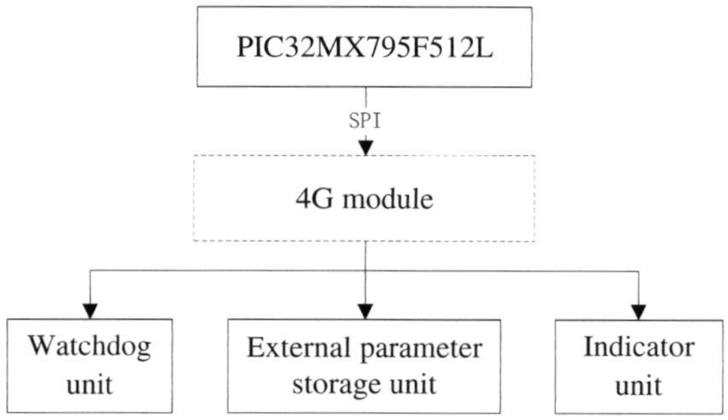

Figure 2.Communication system

Landslide is caused by earthquake, cooling, freezing and thawing, man-made and many other factors. The occurrence of landslides is induced by the role of immediate activities. During the slow movement, the landslide body's displacement is monitored by the simple landslide monitoring system. The monitoring data can be accurate to 0.1mm. Once the landslide displacement value is greater than the warning threshold, the landslide warning device will immediately start transmitting the monitoring data and send an alarm signal to the monitoring center.

Voltage values and external hardware are measured in order to avoid failure of the monitoring equipment if the insufficient power supply or for human or other reasons. The measured voltage is less than the voltage of the normal operation of the instrument or the external device is measured. The monitoring device will send the information to the monitoring center promptly to the staff that the labor cost is reduced to a certain extent, and the staff can't be dispatched for a long time.

2.2. *System software design*

2.2.1. System initialization.
When the installation is completed, the system is initialized. The control center sends instructions to set the internal parameters of the instrument through in the form of SMS, which contains the control information, the independent identification code, time value, timing, displacement warning value, voltage alarm value, displacement initial value, end flag and so on. When the system is set up successfully, the system goes into the low power mode. During system operation, the terminal system receives irregular displacement, voltage, time or monitoring cycle, the control center can also send adjustment instructions to reinitialize the monitoring parameters to ensure normal operation.

System monitoring operation. Data monitoring terminal software as the main data receiving, analysis and processing platform, it receives the site data through the Internet network which comes from wireless sensor network coordinator. The monitoring center needs to receive the collected data stored in the database SQL Server, So that the data can be achieved historical data query and real-time view. You can set the corresponding warning value in the monitoring software. When the scene of a certain collection of data exceeds the warning value, the alarm message will be issued to cause the attention of the monitoring staff. (Figure 3)

IOP Conf. Series: Journal of Physics: Conf. Series **933** (2017) 012017 doi:10.1088/1742-6596/933/1/012017

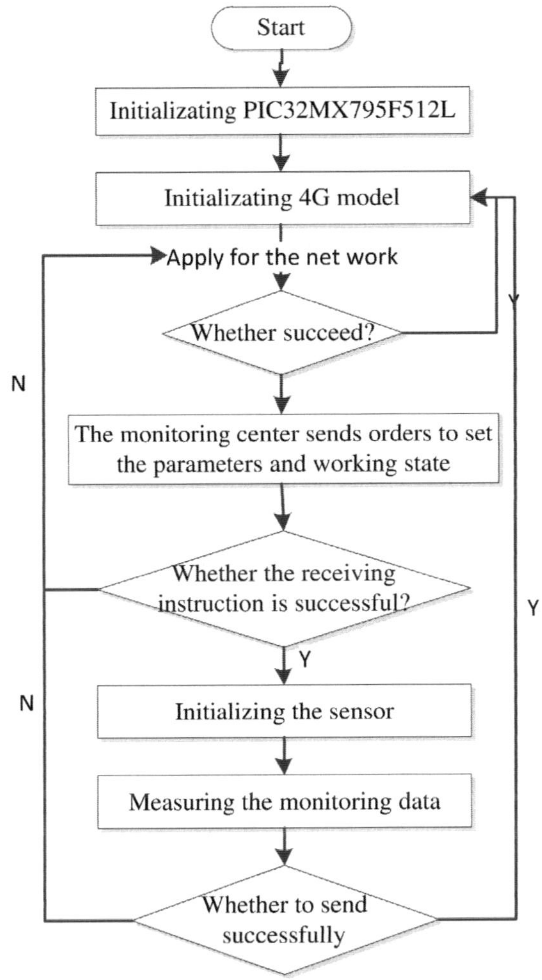

Figure 3Communication system

3. System build implementation

3.1. *High-precision, regular monitoring transmission*

Using micro-power 16-bit AD analog-to-digital conversion module, the way achieves field unattended acquisition displacement data and voltage values. The system is timing to transfer data through the 4G module to the monitoring center on the host. 4G mobile communication network covers a wide area; the signal is strong that making the test data transmission by the distance limit is small, sending high efficiency.

3.2. *Low cost, low power consumption, high reliability*

Choosing low-power components, the system fully considers the low power requirements in the software design process. In a short time, the monitoring task is completed into the dormant state that can save energy. Power supply system is mainly used 12V battery-powered, solar automatic charging system.

3.3. *Easy installation of equipment*

The overall design of the instrument is used in the form of snap card from the connector to the overall support. The installation process is simple and easy to operate. To avoid long-term field operations, the uncertainties of environmental factors affect the field of installation work.

4. Conclusions

Simple landslide monitoring system is to achieve low power consumption, simplicity, serialization, and intelligent that it is more suitable for field unmanned monitoring. The traditional cost-effective PIC32MX795F512L microcontroller is used as a control module.The 4G communication system is used to remote data monitoring. The system uses PIC32MX795F512L internal timer to wake-up microcontroller that it can improve the monitoring accuracy and speed and makes lengthen sleeping time to further reduce the system power consumption. The 4G module connects to the computer terminal through the 4G network, receives the data and sends the instruction. The design of the landslide monitoring system provides more reliable hardware support for long-term field monitoring.

5. References

[1] Zhang Jianli, Xu Xiaozhang, "Composition and analysis of monitoring results of slope movement in geological disastercontrol," Journal of Zhejiang Land and Resources, China, 2009, PP. 54-55.
[2] https://wenku.baidu.com/view/ca8af691daef5ef7ba0d3c7e.html
[3] Wang Jun-dong, Meng Xianshuang, "How to Supervise and Analyze the Slope Moving in Geologic Hazard," Value Engineering, China, 2010, pp. 166.
[4] Huang Zhiyong, "Low power data acquisition system for timing monitoring," Dalian University of Technology, 2003, pp. 8-14.
[5] Sha Fei, "Talking about the Low Power Design of Single Chip Microcomputer," China Computer&Communication, China, 2011, pp. 52-53.
[6] Liu Dong, "Summarization of Monitoring Methods of Landslide Geological Disasters,"Hunan Chemical Geological Engineering Investigation Institute, China, 2009,pp 63-65.
[7] Wu Qingrong, Ding Yuejun, "PIC monolithic integrated circuit low power loss design,"Manufacturing automation, China, 2011, pp.140-142.
[8] Cao Xiuding, Ruan Jun, Zhan Jianshe, Zeng Ke, "Long Distance Real Time Monitoring and Control of Landslideand Data Transmission,"The Chinese Journal of Geological Hazard and Control, China, 2002, pp61-65.

Acknowledgments

I would like to show my deepest gratitude to my supervisor, Dr.Shiyanxin, a respectable, responsible and resourceful scholar, who has provided me with valuable guidance in every stage of the writing of this thesis. Her keen and vigorous academic observation enlightens me not only in this thesis but also in my future study. I would also like to thanks to the project of Land and Resources Public Welfare Industry Science and Technology Special Funds and all my co-workers who have helped me to develop the fundamental and essential academic competence.

IOP Conf. Series: Journal of Physics: Conf. Series **933** (2017) 012018 doi:10.1088/1742-6596/933/1/012018

Cloud Computing: A model Construct of Real-Time Monitoring for Big Dataset Analytics Using Apache Spark

Ameen Alkasem[1], Hongwei Liu[1], Decheng Zuo[1] and Basheer Algarash[1]

[1]School of Computer Science and Technology, Harbin Institute of Technology, Harbin 150001, China

Email: 77ameen@gmail.con

Abstract. The volume of data being collected, analyzed, and stored has exploded in recent years, in particular in relation to the activity on the cloud computing. While large-scale data processing, analysis, storage, and platform model such as cloud computing were previously and currently are increasingly. Today, the major challenge is it address how to monitor and control these massive amounts of data and perform analysis in real-time at scale. The traditional methods and model systems are unable to cope with these quantities of data in real-time. Here we present a new methodology for constructing a model for optimizing the performance of real-time monitoring of big datasets, which includes a machine learning algorithms and Apache Spark Streaming to accomplish fine-grained fault diagnosis and repair of big dataset. As a case study, we use the failure of Virtual Machines (VMs) to start-up. The methodology proposition ensures that the most sensible action is carried out during the procedure of fine-grained monitoring and generates the highest efficacy and cost-saving fault repair through three construction control steps: (I) data collection; (II) analysis engine and (III) decision engine. We found that running this novel methodology can save a considerate amount of time compared to the Hadoop model, without sacrificing the classification accuracy or optimization of performance. The accuracy of the proposed method (92.13%) is an improvement on traditional approaches.

1. Introduction

Large datasets monitored move fast in real-time and tend to be the most valuable as a result [1]. For example, utility cloud service providers may wish to monitor engines to detect faults or anomalies in seconds or quickly engage self-recovery before losing service. To enable these low-latency processing applications, there is a need to design and evaluate a new model to rapidly manage and analyze huge data by means of streaming computation models that scale transparently to large clusters such as Apache Spark [2]. This happens with cheaper applications and low maintenance costs. Among the recent tools and technologies, Apache Spark has become one of the most popular engines for large dataset processing. Evaluating and designing a new model is challenging. Monitoring is an important aspect of systems engineering allowing effective maintenance and evaluation of deployed systems [3]. There is a common set of motivations for monitoring which apply to virtually all areas of computing, including cloud computing; perhaps foremost is capacity planning failure or underperformance

IOP Conf. Series: Journal of Physics: Conf. Series **933** (2017) 012018 doi:10.1088/1742-6596/933/1/012018

detection, redundancy detection, system evaluation and policy violation detection. At their very simplest, monitoring and control entail a three stage process illustrated in Figure 1: (I) The collection of relevant state (II) The analysis of the aggregated state and (III) Decision making as a result of the analysis [4].

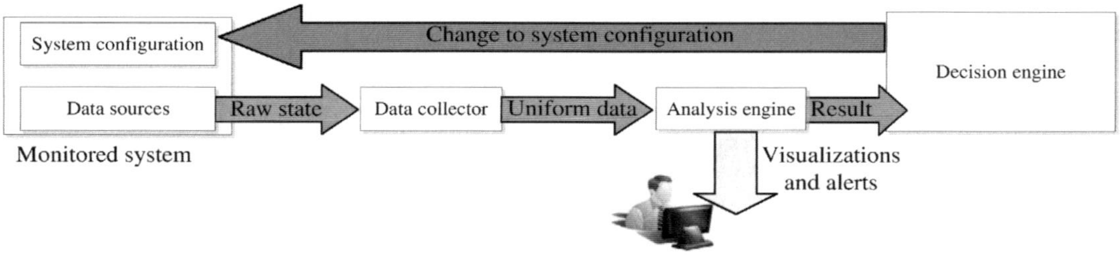

Figure 1. A three stages monitoring process technical

More precisely, professionals use machine learning technologies such as Bayesian Networks (BNs) extensively for the classification and analysis of datasets due to their "learning" capabilities from the combination of datasets. These are often based on training and tests in order to make relatively highly accurate real-time decisions. Apache Spark is a fast general purpose computing engine designed for large-scale data processing. Spark has the advantages of Hadoop of MapReduce; Spark is different from MapReduce job intermediate results can be stored in the memory, thereby eliminating the need to read and write Hadoop Distributed File System (HDFS), so Spark can be better applied to the big dataset and machine learning iterative MapReduce algorithm. Recently, research in this area has abounded. Zhang, et al. [5] provided a flexible architecture called *CloudMonatt* to monitor the security health of customers' VMs within a cloud system. Le et al. [6] developed an efficient multilevel healthy cloud system to analyze data collected from CGU smart clothes using Spark. Aceto et al. [7] surveyed cloud monitoring wherein they identified open issues, main challenges, and future directions in the field. Kumar et al. [8] developed a MapReduce framework for automatic pattern recognition based on fault diagnosis by solving data imbalance problems in cloud-based manufacturing (CBM). Melo et al. [9] developed and proposed a model to evaluate nodes' capacity in a cloud computing environment based on available hardware resources. Nevertheless, comprehensive analysis yields the fact that these researchers, who work in different research areas, did not take into account other effects, such as efficiency, accuracy, and optimization of performance in real-time monitoring of large datasets analysis and classifiers as they pertain to diagnosing faults and anomalies' behaviors.

The contribution of this paper presents a new methodology which includes machine learning algorithms and the Apache Spark to accomplish fine-grained monitoring and control for the fault diagnosis and recovery of the IaaS clouds in real-time. This paper presents a new methodology for constructing a model that optimizes performance of real-time monitoring and improves prediction accuracy. Also, this research performs scalability monitoring to see the speedup of the big datasets processing and analyze this in real-time based on Spark Streaming. The case study here is the failure of virtual machines (VMs) to start up. The methodology proposition ensures that the most sensible action is carried out during the procedure of fine-grained monitoring and generates the highest efficacy and cost-saving fault diagnosis through three construction control steps: (I) Data collection; (II) Analysis engine; and finally (III) Decision engine, as shown in Figure 2. Results show that running the new model across machine learning algorithms and Apache Spark can save a consider able amount of time compared to running the model against a Hadoop without sacrificing the classification accuracy. It can also optimize analytical performance. Results have proven promising, considering an accuracy of 92.13% with an increase in the size of the datasets.

This paper has five sections, beginning with this introduction. In section 2, we describe the tools we used in an empirical evaluation of our methodology. Section 3 focuses on our approach's materials and methods, and explains the analysis, classifier and decision engine models. Section 4 explicates the experimental results, environment and includes a discussion. Finally, we provide our conclusions and future works in Section 5.

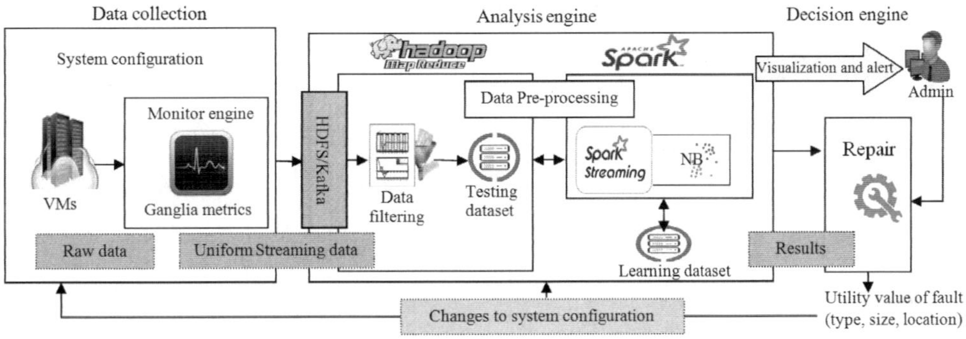

Figure 2. Overview of a new model monitoring workflow based on Apache Spark

2. Tools Used in Empirical Evaluation of Methodology

2.1 Apache Spark Technology

Spark is a distributed computing framework like MapReduce, but it score is a flexible distributed data set that provides a richer model than MapReduce. Specifically, it allows iterations of data sets in memory quickly to support complex data mining algorithms and graph calculation algorithms [3]. It provides a simple programming interface (API's) for various analytics algorithms including real-time data streaming, SQL Queries, graph processing and machine learning. The interface allows for an application developer to easily use the CPU, memory, and storage resources across a cluster of servers for processing large and complex datasets. The advantages of Spark Streaming are:(I) runs on 100+ nodes, and reaches the second delay, (II) memory-based execution engine, with efficient and fault-tolerant features, (III) integration of Spark's batch and interactive queries, and (IV) Up to 10 times faster than Hadoop MapReduce.

Apache Spark is a very simple architecture with only two nodes (Master and Worker) that run with a cluster manager such as Spark, Hadoop or others, as shown in Figure 3

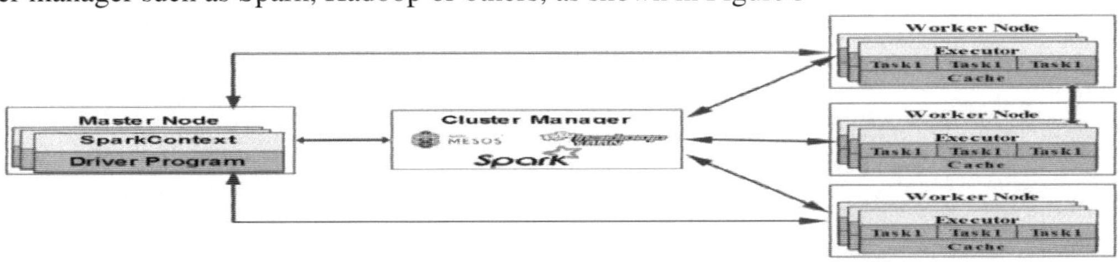

Figure 3. Apache Spark architecture

The Spark program processes the batch by manipulating the interfaces that RDDs (Resilient Distributed Datasets) provide, such as map, reduce, filter and so on. In Spark Streaming, the interfaces provided by the operation of the DStream (RDD sequence representing the data flow) are similar to those the RDD provides.

2.2 Spark Streaming Framework

Spark Streaming is a real-time native computing framework that extends Spark's ability to handle large-scale streaming data. The framework demands data collection from many resources such as TCP sockets, HDFS and Kafka, as well as processing it through complex algorithms expressed with high-level functions such as MapReduce and window [13]. The basic principle of Spark Streaming is to split the input data stream into units of time slices (seconds), and then process each time slice data in a batch-like manner. Figure 4 illustrates the basic principle.

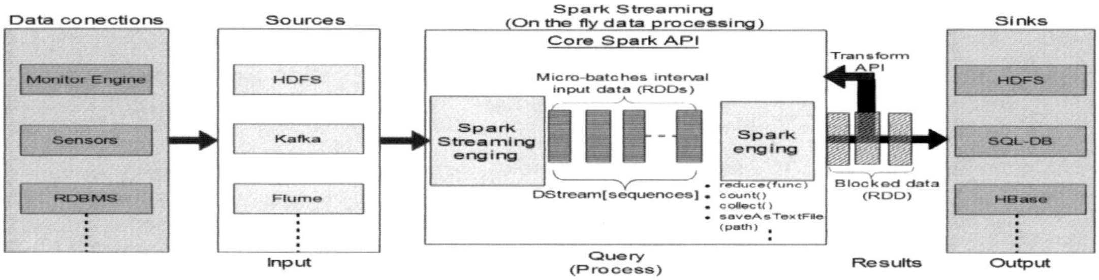

Figure 4. Spark Streaming implementation framework

We used Spark Streaming in this work because it allows us to combine applications such as streaming, batch and interactive queries through rich APIs and memory-based high-speed computing engines. Indeed, it is suitable for applications that require historical data and real-time data combination analysis, though real-time requirements do not require particularly intensive application. In addition, the RDD data reuse mechanism can be more efficient and more fault-tolerant processing [13].

2.3 Structured Streaming

Structured stream processing is another engine based on the Spark SQL optimization engine Catalyst optimizer, ensuring its performance is also strong. This kind of processing ensures data is processed only once, while achieving end-to-end data and high fault tolerance by investigating mechanisms like checkpoints and write ahead logs. The current dataset/DataFrame API supports program languages in Scala, Java, and Python. We can implement flow aggregation, event window, join aggregation, and so on [13].

2.4 Machine Learning with Apache Spark

Machine learning is a technology with strong ties to statistics and optimization; these allow learning from existing data to explore hidden valuable information [14]. It has become one of the most popular techniques for knowledge discovery and predictive analytics, especially with the current exponentially growing data derived from various disciplines such as the medical sciences and business. Many applications, like spam filtering, advertisement targeting, computer vision, and bioinformatics, have adopted machine learning algorithms to better guide leadership's decisions. Apache Spark programming model and the Hadoop platform to analytic and process big datasets in real-time gives a powerful algorithms implementation. Many machine learning algorithms have been investigated to be transformed to the Spark paradigm in order to make use of the HDFS/Kafka resources [15]. The Naïve Bayes classifier (NBC) is one of the supervised learning classification algorithms one can program along the lines of Spark. Currently, the NBC is a scalable machine learning library that supports large dataset processing [16]. Moreover, one can train the NBC very efficiently. Within a single training pass, it computes the conditional probability distribution of each feature given label and then applies Bayes' Equation (1) to compute the conditional probability distribution of a label, given an observation, then employs it for prediction [17].

$$P(c|x) = \frac{P(x|c)P(c)}{P(x)}$$
$$P(c|X) = P(x_1|c) \times P(x_2|c) \times \dots \times P(x_n|c) \times P(c) \quad (1)$$

where

$P(c|x)$: is the posterior probability of the class (target) given predictor (attribute).
$P(c)$: is the class prior probability.
$P(x|c)$: is the likelihood probability.
$P(x)$: is the predictor prior probability (evidence).

3. Our Approach

A new methodology model used the advantages NBC models and Apache Spark Streaming technical. This provided a new approach and framework of monitoring and control of a massive amount of datasets and an analysis in real-time at scale. The new model reduces the cost of the time of a fault diagnosis through fast analytic and engages in classifying large testing and training datasets based on the Spark Streaming technical and HDFS or Kafka storage platforms. The new model monitors a wide range of metrics across the VMs as shown in Figure 5. Most prior work monitored metrics based on the risk levels of consequences according to the symptoms [18].

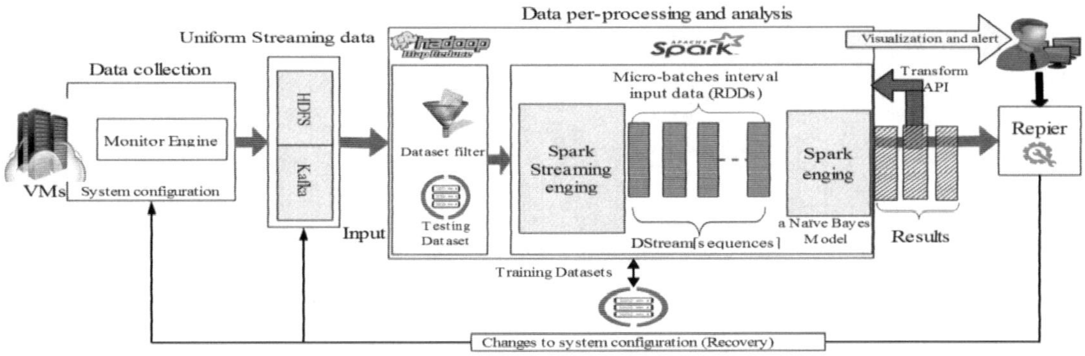

Figure 5. A new methodology proposed framework based on Spark

3.1. The Topology of a New Methodology Proposed

3.1.1. Data Collections for Proposed Model

In this paper, we collected data from large monitoring engines to obtain the metrics of interest (numerical predictors or attributes). Then, we collected the metrics of the VMs by running the VMs on the Xen-hypervisor [19] installed on the host server in combination with pre-processing (reported in Table 1) using Ganglia metrics software [20], as shown in Table 2. Because these metrics exist in the form of numerical data, the numerical variables had to be transformed into their categorical counterparts (binning) before constructing their frequency table to use them as input for the proposed model topology structure. Therefore, as shown in Figure 6, the pre-processing step consists of four steps to translate continuous percentage utilization into interval probability values and generate monitoring vectors of events (M-events) by using our method [21]. We adopted this method with a filtering dataset to remove and process the outlier data and noise using the Extended Kalman Filter (EKF) [22]. As a result, we generated a new method algorithm for transformed a numerical data to binning data (Figure 7).

Table 1. Fault reasons for the start-up failure of VM/Host metrics

Fault category	Fault causes	Measurement Level
CPU utilization	% CPU time used by host CPU during normal sampling period (Host CPU usage) % CPU time during which the CPU of the VM was actively using the physical CPU (VM CPU usage) % CPU time during which the CPU of the VM was ready but could not get scheduled to run on the physical CPU (VM CPU Ready-Time)	VM, Host
Memory usage	% of used memory (Memory usage)	VM, Host
Network overhead	% of network usage (Bandwidth)	VM, Host
I/O storage usage	% of disk usage (Throughput)	Host

Table 2. A sample data collection of CPU-utilization (testing dataset)

Time- Monitoring	VM-CPU-Ready-Time	Host-CPU-usage	VM-CPU-usage
12:02:00 AM	30.00	30.00	30.93
12:07:00 AM	30.00	35.00	78.12
12:12:00 AM	90.00	70.00	92.43
12:17:00 AM	31.84	38.21	42.19
⋮	⋮	⋮	⋮

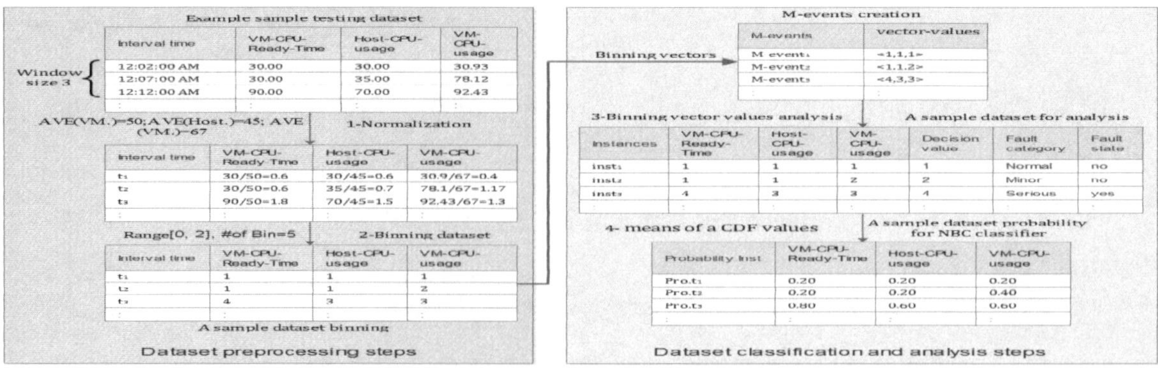

Figure 6. A process method steps and a sample of tables of the datasets transformed

Algorithm 1: Transformed and filtered a data points to data binning

```
Input: Metrics_values[i, j], mean[j], m, r, n, num.
Output: Data binning table.
For i=0 , i<=n; i++ do
     For j=0,j<=num, j++ do
        Normalized[i,j]=Metrices_values[i,j]/mean[i]
          If Normalized [i,j]>n then
              Bin[i,j]=m,
         else
             Bin[i,j]=TRANC(Normlized[i,j]/(r/m)),
    End for
End for
Data_normalized[i,j]=Filtering-dataset(Bin[i,j]),
End Function return(Data_binning).
```

Figure 7. A proposed method algorithms for filtering and transformed a numerical data to data binning

Our method has a buffer size of n (a look-back window) for the metrics of the previously observed n samples (e.g., n=3, range [0, 2] and #Bin=5). The look-back window is used for multiple reasons: (I) shifts of work patterns may render old history data even useless or misleading, (II) at exascale, it is impractical to maintain all history data, and (III) it can be implemented in high speed RAM, which can further increase detection performance. We use a window-size of 3, sampling interval of 5 seconds, and interval length of 14 minutes. The number of data points in the 14 minutes interval are, thus, (14*60)/(3*5) = 56. The metrics in the look-back window at each time instance ($t_1, t_2, \ldots, t_i$, i is number of instances) serve as inputs for the pre-process, as shown in Table 3, so that m-events ($M_1, M_2, \ldots, M_i$) creation are the input for every component metric for our presented methodology, as shown in Table 4. The binning value and the decision value are determined by the following formulas, (2), (3) and (4):

$$\begin{aligned}&\textbf{If}(x > 2)\ \textbf{then}\ \textit{binning value} = 5,\\ &\textbf{else}\ \textit{binning value} = \mathrm{TRUNC}(x/0.4)\end{aligned} \tag{2}$$

$$\text{Decision } \textit{value} = \mathrm{MAX}(\textit{binning value}) \tag{3}$$

Where x is the normalization value for the attributes and 0.4 is a statistic suggested by the probability values. After this, we can begin the classification with a dataset probability (predictor) by means of a cumulative distribution function (CDF) [23] as Table 5 illustrates. For a continuous random variable, the CDF equation is.

$$P(X \le x) = \frac{x - a}{b - a} \tag{4}$$

Where a is the lower limit 0 and b is the upper limit 5, $a \le x \le b$.

Table 3. A sample of data normalization used for binning

	Sample	CPU	Network	Memory
Window size 3	t_1	0.60	0.67	0.46
	t_2	0.60	0.78	1.16
	t_3	1.80	1.56	1.38
	t_4	1.05	1.15	0.97
	⋮	⋮	⋮	⋮

Table 4. A sample of dataset binning with decision values

M-events	CPU	Network	Memory	Decision value
M_1	1	1	1	1

Window size 3	M_2	1	1	2	2
	M_3	4	3	3	4
	M_4	2	2	2	2
	⋮	⋮	⋮	⋮	⋮

Table 5. A sample of the classifier probability (predictor) dataset before training

Process Instance	CPU	Network	Memory	VM/Host state	Fault state
$Pro.inst_1$	0.20	0.20	0.20	Normal	?
$Pro.inst_2$	0.20	0.20	0.40	Minor	?
$Pro.inst_3$	0.80	0.60	0.60	Serious	?
$Pro.inst_4$	0.40	0.40	0.60	Minor	?
⋮	⋮	⋮	⋮	⋮	⋮

3.1.2. Analysis Engine Model

We present a new methodology for large-dataset analysis based on the Hadoop MapReduce and Apache Spark platform, and we implement new algorithms with MapReduce and Spark Streaming to achieve high-performance and efficient classification for analyzing collections of large metrics of test and learn datasets in real time. This section explains the new algorithms and workflow of analysis, as well as the pre-processing method. Figure 8 elucidates the procedure by displaying four new modules to Hadoop along with the techniques we used for method evaluation (Apache Spark):

(1) A monitoring engine: Monitors each VM and host server to collect the metrics of interest (numerical predictors or attributes).
(2) Pre-processing: we test new data from the metrics collection monitor by normalizing it and binning it into intervals, then using the raw time-series data to generate monitoring events (m-events).
(3) Diagnosis engine: Our model analyzes the time series' entropy to locate patterns that signify faults or anomalies in the system being monitoring.
(4) A decision process model: In this model, a look-back window with size 3 creates a dataset table according to the following decision values:

- **If** (0 =decision value < 2) **Then** (fault category is "Normal," node fault state is working "fault=no").
- **If** (2 <decision value =3) **Then** (fault category is "Minor," node fault state is working "fault=no").
- **If** (decision value > 3) **Then** (fault category is "Serious," node fault state is no working "fault=yes").

Our work focuses on enhancing the new proposed methodology model for differing sizes of datasets and assessing the Hadoop MapReduce program and Apache Spark Streaming in having the capability of learning from a past large datasets with the purpose of achieving high-performance fault diagnosis for the metrics of the VMs and host server collected in real time monitoring. Figure 8 demonstrates how we can take raw data from the testing metric data collected by large monitoring engines (Xen-Hypervisor) as shown in Table 9 in Appendix A, and a historical training dataset as shown in Table 10 in Appendix A.

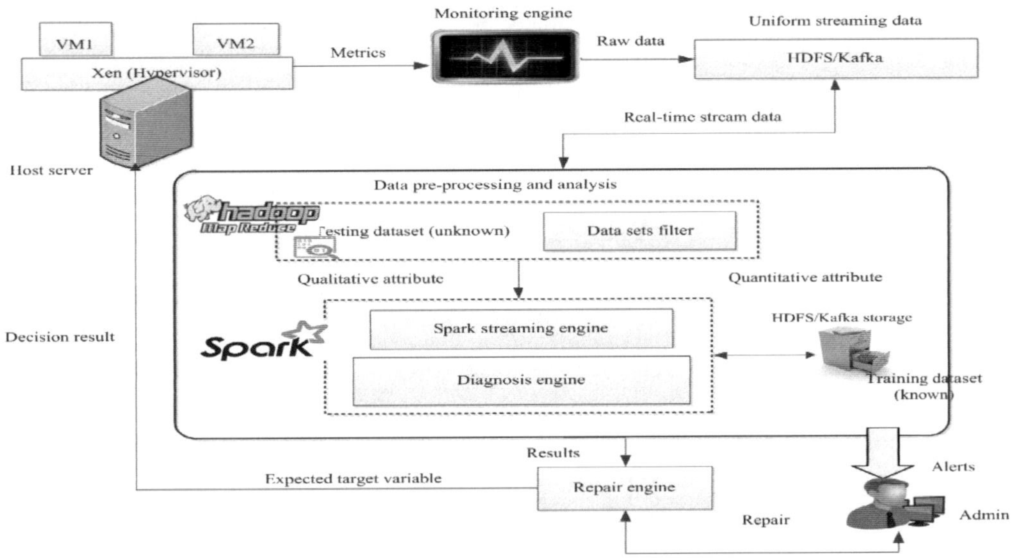

Figure 8. Proposed frameworks for a fine-grained approach with Spark

3.1.3. Diagnosis and Classifier Engine

The proposed diagnosis approach is a general concept representing a combination of analytical tools that all cooperate to form the analysis and diagnosis model (Figure 8). We took our raw data from large engines monitoring a host server, as well as drawing from data research communities have already collected. A monitoring engine collected and processed measurement data such as basic resource metrics for each VM and host server, as listed in Table 1. The diagnosis engine contains the storage of Kafka/HDFS and hybrid intelligent models Hadoop MapReduce and Spark. An HDFS/Kafka is a storage platform for a real-time streaming data collected. Here, we can use the Hadoop to quickly obtain the overall generating testing datasets and input them in the pre-processing and analysis model.

The Spark streaming analyzes m-events diagnosis by pre-processing and the use of I/O pairs to calculate the parameters estimated by an NBC in the Spark engine for large learning datasets. We utilized three statistical measures (recall, precision and accuracy) to evaluate the effectiveness of fault diagnosis in real-time using the Spark streaming engine in Hadoop for large-testing/learning dataset problems; Subsection 3.1.4 and Section 4 unveils more details and results. The data parser first pre-processes all component metrics recorded in a common XML file [24]. Each file contains two parts after pre-processing: (I) The attributes of the components and (II) The dataset, as shown in Figure 9. The model records each observation of component metrics as a single line of datasets where the first three columns represent the metrics of attribute components and the last two columns represent the component state class (normal, minor and serious) and the system fault state class(yes, no). A comma separates each column, which is useful because, by default, Hadoop MapReduce splits the input files by line and passes each line to a mapper function. It then stores all preprocessed component metrics in the master node as a repository while waiting for further sampling. The proposed approach maintains a buffer size of 3 (a look-back window) of the last n samples' metrics.

The metrics observed in the look-back window at each time instance serve as inputs to be pre-processed.

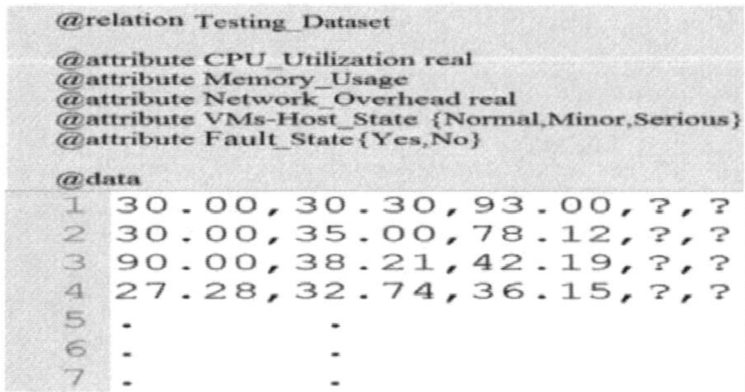

```
@relation Testing_Dataset

@attribute CPU_Utilization real
@attribute Memory_Usage
@attribute Network_Overhead real
@attribute VMs-Host_State {Normal,Minor,Serious}
@attribute Fault_State{Yes,No}

@data
1 30.00,30.30,93.00,?,?
2 30.00,35.00,78.12,?,?
3 90.00,38.21,42.19,?,?
4 27.28,32.74,36.15,?,?
5 .        .
6 .        .
7 .        .
```

(a) Testing dataset (unknown)

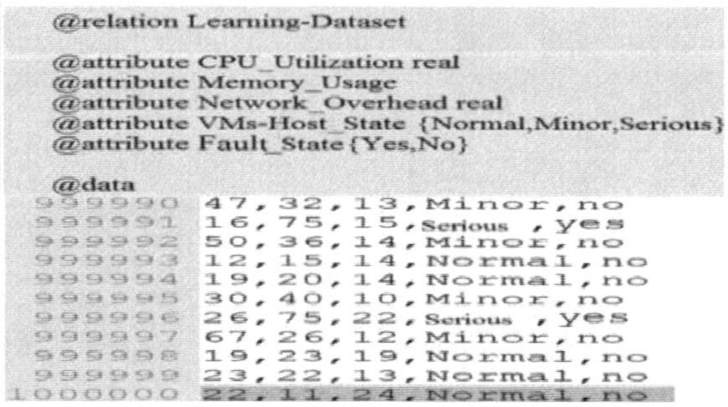

```
@relation Learning-Dataset

@attribute CPU_Utilization real
@attribute Memory_Usage
@attribute Network_Overhead real
@attribute VMs-Host_State {Normal,Minor,Serious}
@attribute Fault_State{Yes,No}

@data
999990 47,32,13,Minor,no
999991 16,75,15,Serious ,yes
999992 50,36,14,Minor,no
999993 12,15,14,Normal,no
999994 19,20,14,Normal,no
999995 30,40,10,Minor,no
999996 26,75,22,Serious ,yes
999997 67,26,12,Minor,no
999998 19,23,19,Normal,no
999999 23,22,13,Normal,no
1000000 22,11,24,Normal,no
```

(b) Training dataset (known)

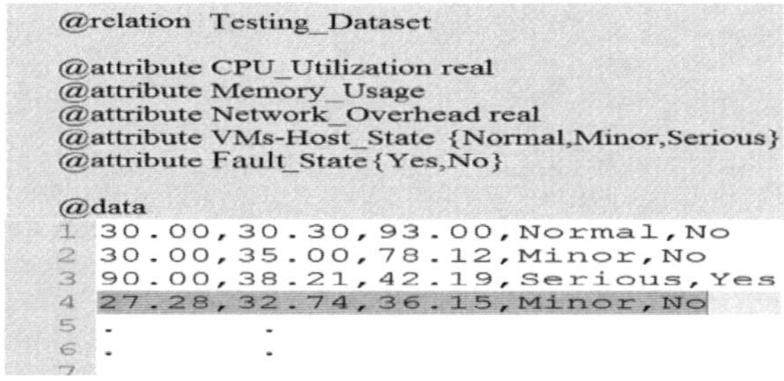

```
@relation Testing_Dataset

@attribute CPU_Utilization real
@attribute Memory_Usage
@attribute Network_Overhead real
@attribute VMs-Host_State {Normal,Minor,Serious}
@attribute Fault_State{Yes,No}

@data
1 30.00,30.30,93.00,Normal,No
2 30.00,35.00,78.12,Minor,No
3 90.00,38.21,42.19,Serious,Yes
4 27.28,32.74,36.15,Minor,No
5 .        .
6 .        .
7
```

(c) Testing dataset after classification and learning (known)

Figure 9. Datasets modeling (XML format)

3.1.4. Decision Engine Model

The classification of component states and system states is a key step in workflow. Figure 8 shows the job sequence of this step. When the real-time test dataset is ready in HDFS/Kafka, a new model starts the training job to build a model. The algorithm combines the test data with the model and generates an intermediate table, then classifies the job and simultaneously computes the probability of each component in the three classes. Finally, it makes a decision about the final system state and records statistics regarding the contingency table and intermediate values. In our experiments, we used two datasets: (I) New testing datasets (unknown) and, (II) Training datasets (known). Tables 6 and 7 show how these data sets work by converting to XML format for input to the proposed model algorithms. An example of the XML files formation (Figure 9(a)-(b)) is our simple dataset framework for a state-based probability model that predicts component utilization. In Figure 10, the measurement level varies from 0 to 100%. The component utilization uses CPU ϵ {0–25%, 26–75%, 76–100%} as thresholds. We observed the percentage component utilization at discrete times $t_1, \dots, t_n$.

The new dataset holds the model classifier results by combining the test and training datasets (Figure 9(c)). In the process, we first define our task to classify the three state components using a model proposed. The system cannot work if one component is faulty. We use 0, 1, and 2 to represent the system and component states, where 0 denotes good status (normal working conditions), 1 denotes a minor fault and 2 denotes a serious fault. For example, CPU utilization, memory usage, and network represent three basic components while host state server represents the system state (Figure 10). This results in only three classes (normal, minor and serious) for the component state measurements and only two (yes, no) for the system fault state [25]. To simplify the problem, we choose the same number of normal, minor, and serious measures for the components. We must then convert the classification problem to a counting problem on the training datasets and test datasets.

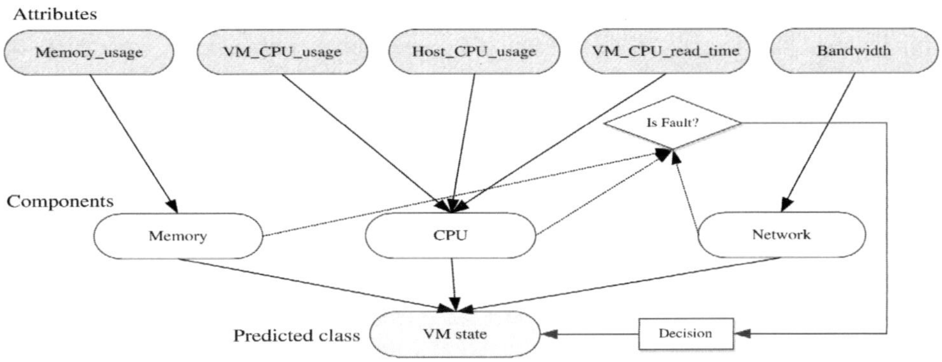

Figure 10. Model predicting component utilization for VM

We divided the problem into two procedure programs:

1) The first is a combination of two java and Scala programs for monitor and control of big dataset training, filtering, and streaming datasets (Figure 16 in Appendix B). All training and testing instances are fed into these programs to produce a model for all unique attributes and filtering datasets with their frequencies of normal, minor and serious component states. The model also contains information necessary for the final classification in real-time processes based on Apache streaming engine.
2) The second program regards classifying and analyzing the test and training dataset, as show in Figure 17 in Appendix B. In this program, the training and the testing dataset measures combine into an intermediate table with all information necessary for the final classification. This program classifies all instances for the test dataset and gets the final results of fault state classes (yes, no). Moreover, it writes final resulting classification to HDFS or Kafka storage platforms.

After the two programs finish, the results collector retrieves the model classification results intermediate values table and test data statistics from HDFS/Kafka storage. By the end of these programs, instances have been classified and evaluation into a yes/no final fault state with normal, minor and serious state classes, which one can view in Table 6 and Figure 11.

Table 6. A sample of the dataset classifier probability after training

Process. Instance	CPU	Network	Memory	VMs/Host state	Fault state
Process.inst$_1$	0.25	0.25	0.25	**Normal**	**no**
Process.inst$_2$	0.25	0.25	0.50	**Normal**	**no**
Process.inst$_3$	1.00	0.75	0.75	**Serious**	**yes**
Process.inst$_4$	0.50	0.50	0.50	**Minor**	**no**
Process.inst$_5$	0.40	0.31	0.60	**Minor**	**no**
⋮	⋮	⋮	⋮	⋮	⋮

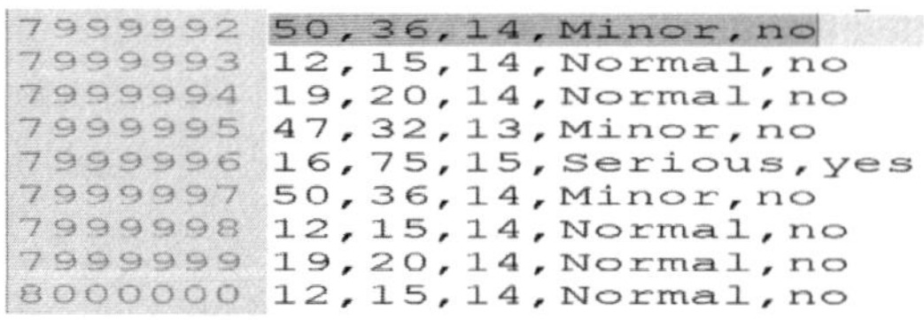

```
7999992 50,36,14,Minor,no
7999993 12,15,14,Normal,no
7999994 19,20,14,Normal,no
7999995 47,32,13,Minor,no
7999996 16,75,15,Serious,yes
7999997 50,36,14,Minor,no
7999998 12,15,14,Normal,no
7999999 19,20,14,Normal,no
8000000 12,15,14,Normal,no
```

(a) Testing dataset after diagnosis and analysis

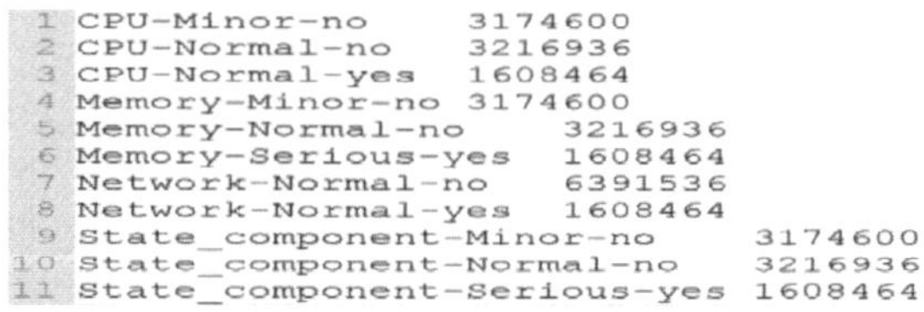

```
1 CPU-Minor-no 3174600
2 CPU-Normal-no 3216936
3 CPU-Normal-yes 1608464
4 Memory-Minor-no 3174600
5 Memory-Normal-no 3216936
6 Memory-Serious-yes 1608464
7 Network-Normal-no 6391536
8 Network-Normal-yes 1608464
9 State_component-Minor-no 3174600
10 State_component-Normal-no 3216936
11 State_component-Serious-yes 1608464
```

(b) Spark streaming result for big dataset testing

Figure 11. Implementation results of proposed approach with new programs algorithms

4. Experimental Results

4.1. Experiment Environment

For experimental purposes, we implemented our proposed model's programs using Scale and Java coding. All programs ran on several Apache Spark clusters with different number of nodes. The experiment setup used 2 VMs (VM1 and VM2) on a Xen-Hypervisor platform hosted on one Dell blade server. Apache Spark cluster formed through a local server machine service. Each node inside the cluster has the following configuration: vCPU (3.9 GHz), 1 TB storage and 64 GB RAM. The dataset contains multi instances stored inside local server machines by HDFS/Kafka storage management then read through the master node. The architecture of the Apache Spark cluster appears in Figure 12. For comparison, we have also implemented the same environment using Hadoop platform without Spark.

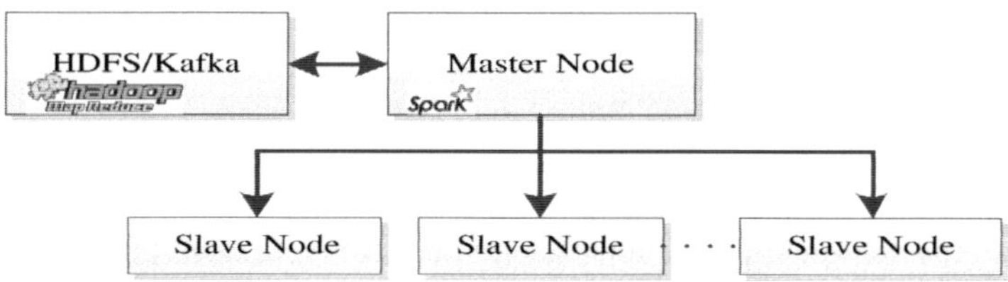

Figure 12. Architecture of the experiment of Apache Spark's cluster

4.2. Evaluation with Real-time Monitoring

We implemented our approach at both the VMs level and the host server level, so the local time series is calculated first for the VMs. Their aggregation for the host server's global time series follows. The first implementation uses Xen-Hypervisor and the Ganglia metrics method to record and identify global resource provisioning anomalies [26]. We injected forty anomaly samples into the testbed, leading to global resource consumption by the anomalies/faults without excluding the CPU-utilization of the running host server (Figure 13). We collected the VM and host metrics using the Ganglia metrics method and analyzed them in a fault detector and classifier. To test the scalability and performance of the new model, we varied the size of the datasets size from 8,000,000 to more than 64,000,000 instances of recorded metrics in each class with datasets between 1.26 GB and 12 GB or over.

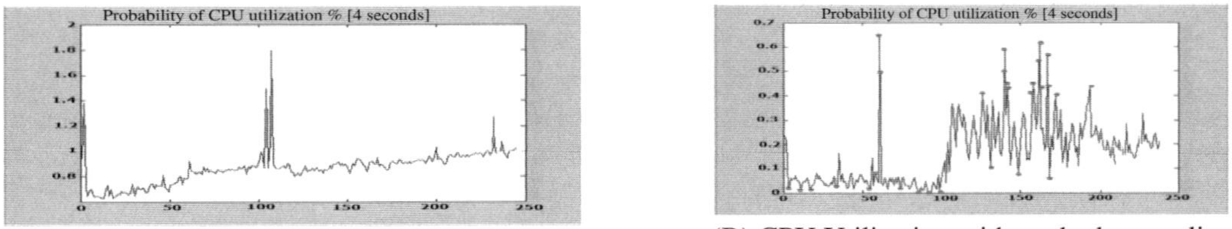

(A) CPU Utilization availability (B) CPU Utilization with testbed anomalies

Figure 13. CPU Parameters using Ganglia monitored metrics

4.3. Results

The resulting statistics include the classification frequency and accuracy of CPU utilizations for two VMs and host server machines by programs algorithms (Figure 14). These results demonstrate a high accuracy for the classification of component states (normal, minor and serious). We show the accuracy and throughput of the system evaluation by Hadoop and Apache Spark in Figure 15.

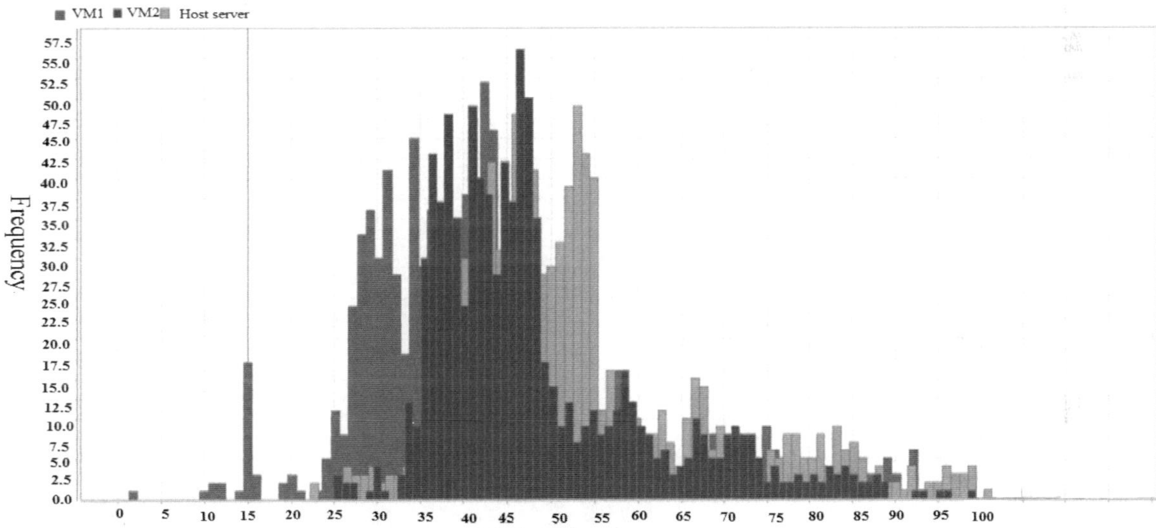

Figure 14. Histogram frequency of CPU utilizations running on VM1, VM2 and Host server

4.4. Discussion

Equations in Table 7's four statistical measures [27][28] helped us evaluate the effectiveness of the construction steps of the monitor procedure, thereby allowing us to detect the faults in the testbed of our experimental setup. We explain this further in Section 4.1. Our experimental results reveal several interesting findings of the evaluation of NBC in our model with Apache Spar. We also committed to a performance summary and results of different implementations without the use of Spark (Figure 15). We achieved high accuracy of up to 92.13% and a 3–15% false alarm rate. The use of Apache Spark Streaming can be optimized to speed up the parameter learning stage in classification and analysis for a wide range of input and dataset sizes even very large ones. The number of instances the system can process and analyze in one second increased from 1.20 GB to more than 12 GB. Table 8 presents a performance summary and various results.

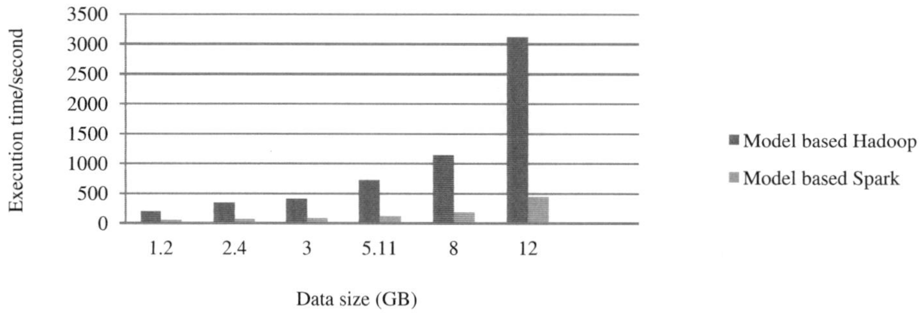

Figure 15. Throughput of system of different model implementations based on Hadoop and Spark

Table 7. Four measures in statistics

Precision	Recall	Accuracy	False-alarm rate (FAR)
$\frac{successfuldetections}{oftotalalarms}$	$\frac{successfuldetections}{oftotalanomalies}$	$\frac{2 * precision * recall}{precision + recall}$	$1 - presision$

Table 8. Performance summary results

Model	Accuracy (%)	Time(Second)*10^3	Dataset size(GB)
Proposed Model with Hadoop	72.86	3.210	12
A new Proposed Model with Apache Spark	**92.13**	**0.450**	12

Finally, our new approach methodology using Apache Streaming can scale over 64 million instances of recorded metrics in each class from a monitoring engine with a high accuracy and saving cost of the time compared to other models.

5. Conclusion

In this paper, we presented a new fine-grained fault-tolerance platform for monitoring and control methodology of big datasets in real-time based on Spark Streaming. Our methodology proposition ensures that the most sensible action occurs during the procedure of fine-grained monitoring, generating the highest possible efficacy and cost-saving fault diagnosis through three construction control steps: (I) data collection, (II) an analysis engine, and (III) a decision engine via Apache Spark. The additional modules inserted to evaluate the newly proposed algorithms resulted in good classification with a high accuracy of up to 92.13% with a 3–15% false alarm rate. Based on the experimental results, a high-performance evaluation of real-time monitoring and control for large datasets using Apache Spark has a better average running time than when using Hadoop. When the size of big datasets is increases, the results on the Apache Spark cluster has good scalability and has better-than-average running time. The key ideas in future work, we have planning to extend our work to develop and design a new model of performance testing for shared dynamic cloud services. The model will be achieves end-to-end performance testing management framework that can troubleshoot, analysis, classify and recovery actions for virtualized cloud based fault bottlenecks and anomalies behavior.

IOP Conf. Series: Journal of Physics: Conf. Series **933** (2017) 012018 doi:10.1088/1742-6596/933/1/012018

Appendix A. A sample datasets

Table 9. A sample dataset of testing metrics before training and classification

Time_monitoring	CPU_utilization	Network_overhead	Memory_usage	VM/Host _state	Fault state
12:02:00 AM	30.00	30.00	30.93	?	?
12:07:00 AM	30.00	35.00	78.12	?	?
12:12:00 AM	90.00	70.00	92.43	?	?
12:17:00 AM	31.84	38.21	42.19	?	?
12:22:00 AM	27.28	32.74	36.15	?	?
12:27:00 AM	32.1	38.52	42.53	?	?
12:32:00 AM	27.42	32.9	36.33	?	?
12:37:00 AM	32.1	38.52	42.53	?	?
12:52:00 AM	65.54	78.65	86.84	?	?
⋮	⋮	⋮	⋮	⋮	⋮

Table 10. A sample historical training dataset

Time_ monitoring	CPU_utilization	Network_overhead	Memory_usage	VM/Host _state	Fault state
10:12:00 AM	57.22	68.66	99.05	Serious	yes
10:12:05 AM	23.34	28.01	35.02	Normal	no
10:17:00 AM	58.96	70.75	20.12	Minor	no
10:19:02 AM	55.08	66.10	72.98	Minor	no
10:20:25 AM	32.16	38.59	42.61	Normal	no
10:21:03 AM	44.86	53.83	59.44	Minor	no
10:22:20 AM	31.84	38.21	42.19	Normal	no
10:27:70 AM	27.28	32.74	36.15	Normal	no
10:28:11 AM	61.16	73.39	81.04	Serious	yes
10:30:32 AM	38.1	45.72	50.48	Minor	no
⋮	⋮	⋮	⋮	⋮	⋮

Appendix B. A sample java and Scala programs coding

```
public class Training_Filtering_Dataset {
    public static class JobMapper extends Mapper<Object, Text, Text, IntWritable>{
        private final static IntWritable one = new IntWritable(1);
        private Text word = new Text();
        public void map(Object key, Text value, Context context) throws IOException, InterruptedException {
         String component_name = "";
         String[] row =   value.toString().split(",");
         if(row.length >= 5){
              String state_component=row[3];
              for (int col=0; col<=3; col++) //columns 0-2 contain data about what the components measures
                {
                   if (col==3 ){
                   state_component=row[3];
                   component_name="State_component";
                   }else if (Double.parseDouble(row[col])>=0 && Double.parseDouble(row[col])<=25){
                   state_component="Normal";
```

```
                }else if (Double.parseDouble(row[col])>=26 && Double.parseDouble(row[col])<=74){
                    state_component="Minor";
                }else if (Double.parseDouble(row[col])<=100 && Double.parseDouble(row[col])>=75){
                    state_component="Serious";}
                    if (col==0)
                    component_name="CPU";
                else if (col==1)
                    component_name="Memory";
                else if(col==2)
                    component_name="Network";
                    word.set(component_name+"-"+state_component+"-"+row[4]);
                    context.write(word, one);
        }}}
public static class JobReducer
            extends Reducer<Text,IntWritable,Text,IntWritable> {
        private IntWritable result = new IntWritable();
        public void reduce(Text key, Iterable<IntWritable> values, Context context) throws IOException,
InterruptedException {
            int sum = 0;
            for (IntWritable val : values) {
                sum += val.get();}
            result.set(sum);
            context.write(key, result);
        } }
// Scale program for engine
    package streaming
      import org.apache.spark.streaming.{Seconds, StreamingContext}
      object DataStream {
      def main(args: Array[String]) {
      val ssc = new StreamingContext("local[4]", "DataStream", Seconds(3))
      val test_dataset = ssc.textFileStream("hdfs://localhost:9000/result")
      val instances = test_dataset.flatMap(_.split(" "))
      val pairs = instances.map(word => (word, 1))
      val instancesCounts = pairs.reduceByKey(_ + _)
      instancesCounts.print()
     ssc.start()
     ssc.awaitTermination()
  }}
```

Figure 16. A programs Java and Scale coding for training, filtering and streaming dataset

```
public class Classify {
public static void main(String[] args) throws Exception {
    BufferedReader breader=null;
    breader= new BufferedReader(new FileReader("("hdfs://127.0.0.1:9000/Input_dataset-_train"));
    Instances train=new Instances(breader);
    train.setClassIndex(train.numAttributes()-1);
    breader=new BufferedReader(new FileReader("hdfs://127.0.0.1:9000/Input_dataset-_test"));
    Instances test=new Instances (breader);
    test.setClassIndex(train.numAttributes()-1);
    breader.close();
    J48 tree=new J48();
    tree.buildClassifier(train); //build classifier
    Instances labeled= new Instances(test);
```

```
            for (int i=0;i<test.numInstances();i++){ // label instances
            double clsLabel=tree.classifyInstance(test.instance(i));
            labeled.instance(i).setClassValue(clsLabel); }
      BufferedWriter writer=new BufferedWriter(new FileWriter("hdfs://127.0.0.1:9000/out_
spark/out_new_dataset_test.xml "));
      writer.write(labeled.toString());
      writer.close();
      StartNBC.main(args);}}
public class StartNBC{
public static void main(String[] args )throws Exception{
BufferedReader breader =null;
breader =new BufferedReader(new FileReader("c:/algorithms/ result_classify3.xml"));
Instances train=new Instances (breader);
train.setClassIndex(train.numAttributes()-1);
breader.close();
NaiveBayes nB=new NaiveBayes();
nB.buildClassifier(train);
Evaluation eval=new Evaluation(train);
eval.crossValidateModel(nB, train, 20, new Random(1), args);
System.out.print(eval.toSummaryString("\nReaults\n======\n", true));
System.out.println("The F.Measure= "+eval.fMeasure(1)+"%"+"      "+ "The Precision="+eval.precision(1)+"
"+"%"+ "The Recall="+eval.recall(1)+"%"); }}
```

Figure 17. A program coding for combining the testing and training datasets classifier and evaluation results

6. References

[1] Witten, I. H., Frank, E., Hall, M. A., & Pal, C. J., "Data Mining: Practical machine learning tools and techniques," Morgan Kaufmann, 2016.

[2] Alsheikh, Mohammad Abu, et al., "Mobile big data analytics using deep learning and apache spark," IEEE network, vol. 30 no. 3, pp. 22-29, 2016.

[3] Agrawal, Dakshi, and Vasileios Pappas,"Virtual machine monitoring in cloud infrastructures," U.S. Patent, no. 9, 600,308, 21 Mar. 2017.

[4] Peng, X., Tang, Y., Du, W., & Qian, F. ,"Multimode Process Monitoring and Fault Detection: A Sparse Modeling and Dictionary Learning Method," IEEE Transactions on Industrial Electronics, vol. 64, no.6, pp. 4866-4875, 2017.

[5] Zhang, Tianwei, and Ruby B. Lee. ,"Monitoring and Attestation of Virtual Machine Security Health in Cloud Computing," IEEE Micro, vol. 36, no. 5, pp. 28-37, 2016.

[6] Le, M. K., Chang, H. T., Chang, Y. M., Hu, Y. H., & Chen, H. T., "An efficient multilevel healthy cloud system using Spark for smart clothes," In Computer Symposium (ICS), 2016 International (pp. 182-186). IEEE, December 2016.

[7] Aceto, G., Botta, A., De Donato, W., & Pescapè, A., "Cloud monitoring: A survey. Computer Networks," vol. 57 no. 9, pp. 2093-2115, 2013.

[8] Ajay Kumar, Ravi Shankar, Alok Choudhary & Lakshman S. Thakur, "A big data MapReduce framework for fault diagnosis in cloud-based manufacturing," International Journal of Production Research, vol. 54, no. 23, pp. 7060-7073, 2016.

[9] Melo, C., Matos, R., Dantas, J., & Maciel, P., "Capacity-Oriented Availability Model for

Resources Estimation on Private Cloud Infrastructure." In Dependable Computing (PRDC), 2017 IEEE 22nd Pacific Rim International Symposium on (pp. 255-260), IEEE, January - 2017.

[10] Zaharia, Matei, et al., "Apache Spark: A unified engine for big data processing," Communications of the ACM, vol. 59, no. 11, pp. 56-65, 2016.

[11] Shyam, R., HB, B. G., Kumar, S., Poornachandran, P., & Soman, K. P. , "Apache spark a big data analytics platform for smart grid," Procedia Technology, vol. 21, pp. 171-178, 2015.

[12] Meng, X., Bradley, J., Yavuz, B., Sparks, E., Venkataraman, S., Liu, D., ... & Xin, D., "Mllib: Machine learning in apache spark," The Journal of Machine Learning Research, vol. 17, no.1, pp. 1235-1241, 2016.

[13] Ranjan, Rajiv, "Streaming big data processing in datacenter clouds," IEEE Cloud Computing, vol. 1, no. 1, pp. 78-83, 2014.

[14] Kotsiantis, Sotiris B., I. Zaharakis, and P. Pintelas., "Supervised machine learning: A review of classification techniques," pp. 3-24, 2007.

[15] Gupta, "A. Learning Apache Mahout Classification," Packt Publishing Ltd; Feb - 2015.

[16] Box, George EP, Gwilym M. Jenkins, Gregory C. Reinsel, and Greta M. Ljung, "Time series analysis: forecasting and control," John Wiley & Sons, 2015.

[17] Krishnan, Venkatarama, and Kavitha Chandra, "Probability and random processes," John Wiley & Sons, 2015.

[18] Ameen Alkasem, Hongwei Liu, and DechengZuo, "Utility Cloud: A Novel Approach for Diagnosis and Self-healing Based on the Uncertainty in Anomalous Metrics," In Proceedings of the 2017 International Conference on Management Engineering, Software Engineering and Service Sciences (ICMSS '17), Yulin Wang (Ed.), pp. 99-107, ACM, New York, NY, USA, 2017.

[19] Chisnall, David, "The definitive guide to the xen hypervisor," Pearson Education, 2008.

[20] Massie, Matthew L., Brent N. Chun, and David E. Culler, "The ganglia distributed monitoring system: design, implementation, and experience," Parallel Computing, vol. 30, no. 7, pp. 817-840 2004.

[21] Wang, Chengwei, Vanish Talwar, Karsten Schwan, and ParthasarathyRanganathan, "Online detection of utility cloud anomalies using metric distributions," In Network Operations and Management Symposium (NOMS), 2010 IEEE, pp. 96-103. IEEE, 2010.

[22] Fujii, Keisuke, "Extended kalman filter," Refernce Manual, 2013.

[23] Doane, David P., and Lori E. Seward, "Applied statistics in business and economics," USA: Irwin, 2005.

[24] Gawrilow, Ewgenij, Simon Hampe, and Michael Joswig, "The polymake XML file format," In International Congress on Mathematical Software, pp. 403-410. Springer International Publishing, 2016.

[25] Alkasem A, Liu H, Decheng Z, Zhao Y. "AFDI: A Virtualization-based Accelerated Fault Diagnosis Innovation for High Availability Computing. arXiv preprint arXiv:1507.08036., Jul 2015.

[26] A Ameen, H Liu, M Shafiq , and D Zuo, "A New Theoretical Approach: A Model Construct for Fault Troubleshooting in Cloud Computing," Mobile Information Systems, Article ID

9038634, 6 September 2017.

[27] Mertler, Craig A., and Rachel Vannatta Reinhart, "Advanced and multivariate statistical methods: Practical application and interpretation," Routledge, 2016.

[28] Alkasem, Ameen, and Hongwei Liu, "A Survey of Fault-tolerance in Cloud Computing: Concepts and Practice," Research Journal of Applied Sciences, Engineering and Technology, vol. 11, no. 12, pp. 1365-1377, 2015.

Acknowledgment

We are also thankful to anonymous reviewers for their valuable feedback and comments for improving the quality of the manuscript.

IOP Conf. Series: Journal of Physics: Conf. Series **933** (2017) 012019 doi:10.1088/1742-6596/933/1/012019

The Use of Vocational Guidance E-Systems in Colleges of Technology, Saudi Arabia

Saad Alsaidan; Li Zhang

School of Management, Harbin Institute of Technology, 150001, China

Corresponding author: s_alsaidan@hotmail.com

Abstract. This study aimed to explore how much interest do technical colleges students have in vocational guidance E-Systems and to detect the most important E-Systems of vocational guidance that they use. The researcher collected data, through questionnaires, from 311 students in their fourth, fifth, or sixth semesters in the colleges of technology in Saudi Arabia. The data were analysed using the chi square test, factor analysis, and descriptive analysis. This study shows the importance in the integration of the E-Systems to vocational guidance. The model includes: the use of e-vocational consulting programs, e-extracurricular programs, e-vocational activities, informational systems, and e-exams for vocational tendency, which were rarely studied before. The results indicated that the college student is highly motivated by vocational guidance using the E-system, unlike the traditional methods of vocational guidance, and young students are motivated by the E-system when it is used in vocational guidance. The study conclude that the integration of the E-system has significantly contributed to vocational guidance. The study recommends that the e-system be implemented to all areas of decision-making. Practitioners should Interact with students through the E-Systems and arrive at decisions that are good for educational growth.

Key words: E-Systems, decision making, vocational guidance, and Technology.

1. Introduction

Recently, vocational education has integrated the use of ICT to support vocational guidance. In the current economy, many of those exiting school and college spend a long time in the labor market before they can secure a long-term job. The average age of beginning apprentices in Saudi Arabia is 28 years old. TVTC practitioners and career guidance counselors should work together closely to monitor labor market conditions and requirements in order to provide optimal services to students. Since a large proportion of students exiting secondary school will not attend universities, all stakeholders need to work together to improve the image of TVTC so that parents and students can appreciate the relevance and importance of vocational training.

The EU acknowledged that the individual is the most important agent for constructing his or her qualifications by upgrading, combining, and using acquired knowledge, skills, and training to make themselves employable and adaptable in our changing world. However, in the present environment, where choosing and preparing for a career is like attempting to shoot at a moving target, proper career guidance and counseling are of critical importance to ensure seamless learning throughout life. The purpose of vocational guidance and counseling is to help students to: "understand and appreciate their talents; relate effectively to others; explore career alternatives; develop appropriate educational and vocational training plans; implement and complete their plans; and integrate successfully in society and the labor market" [1].

Saudi Arabia has developed national career development guidelines to assist educators in implementing career development programs in elementary, secondary, and post-secondary education and training. The model includes: the use of e-vocational consulting programs, e-extracurricular programs, e-vocational activities, informational systems, and e-exams for vocational tendency.

Career assessment provided over the Internet is just one example of interventions that can be problematic because they exclude the help of a counselor. In some high schools, career development activities consist solely of computer-assisted career guidance programs. Fraenkel J. and Wallen N. found that counselor-free interventions were not as effective as interventions that involved a counselor (e.g., individual counseling, workshops, group counseling) [2]. They also found that people who used a vocational computer system supplemented by guidance had better results than those just used a computer system[3].

Some analytics regard the proliferation of counselor-free interventions, both through the Internet and in other settings, as a threat to the career counseling field and a disservice to clients. Unless career counselors are more active in informing others of the efficacy of career counseling, administrators and organizational decision-makers may see counselor-free services as a less expensive alternative to career counseling [4].

In essence, computer technologies significantly modify the techniques of providing vocational services and provide an alternative route for offering vocational services. What, then, is the role of vocational guides? Are they really neglected? How can vocational guides take advantage of advanced technologies to enhance vocational guidance services?

Technology must become vocational guidance' best assistant, not their competitor. Computers will fulfill much of the ponderous and labor-intense work, such as saving files, prospecting for data, managing data, and looking for information. Discussion and deliberation regarding cases can be done using Different technology modes. I recognize that technology will enhance vocational guides' performance, but not replacing their tasks. Computer technologies cannot subrogate for the guide's role in facilitating self-conscious, self-discovery, and the building of an individual's orientation in his vocational life.

Vocational guides, being exempted from routine work by technology, can really use what they are trained for: serving people to make meaning of their life and to develop adaptive skills to regulate well to their environment. Vocational guides do not just match people with their jobs, but also help people to enhance their self-efficiency to work and life.

This study aims to explore how much interest do technical colleges students have in vocational guidance E-Systems and to detect the most important E-Systems of vocational guidance that they use.

2. METHODOLOGY

2.1. Research Questions

This study answered the following three research questions:

1. How much interest do students have in vocational guidance E-Systems?
2. What are the most important E-Systems of vocational guidance that students use?
3. How do demographics affect students' interest towards vocational guidance E-Systems?

2.2. Sample

The research population consists of students in the colleges of technology in Saudi Arabia. The oldest five colleges were selected from the Northern Region in Saudi Arabia by geographic region, including the colleges of technology in Hail, Tabuk, Al-Jouf, Arar, and Quraiyat.

The participants consisted of (311) students in their fourth, fifth, and sixth semesters in the colleges of technology in Saudi Arabia. The reason for choosing the students from the fourth semester onwards is that the students of the fourth semester and onwards would have had enough exposure to vocational guidance E-Systems. The students from within the fourth, fifth, and sixth semesters were then chosen on the basis of simple random sampling.

2.3. Instrument
A questionnaire was developed to collect responses from the students. The questionnaire consisted of two sections: one section consisted of demographic variables (age, college of technology, and study major) and the other section consisted of 7 test items.

The test item statements were the following:

- E-vocational consulting programs
- E-extracurricular programs
- E-vocational activities
- Information systems
- Computer labs
- Knowledge platforms
- E-exams for vocational tendency

The students were participated by expressing their level of agreement or otherwise on a five-point Likert Scale: Strongly Agree (SA), Agree (A), Undecided (UD), Disagree (D), or Strongly Disagree (SD). The experts validated the questionnaire. In order to measure the consistency of the questionnaire, a pilot study was conducted on 311 students of the oldest five colleges in the northern area of Saudi Arabia. The Cronbach's alpha value was 0.86.

2.4. Analysis
The data were analyzed using the chi square test, factor analysis, and descriptive analysis. SPSS version 20 and AMOS were used in the study.

3. RESULTS

Table 1. Communalities

	INITIAL	EXTRACTION
I use an e-vocational consulting programs	.319	.366
I use an e-extracurricular programs	.328	.375
I use e-vocational activities	.214	.251
I use an information systems	.283	.336
I use a computer labs to acquire knowledge	.340	.445
I use knowledge platforms	.298	.346
I use an e-exams for our vocational tendency	.183	.203
Extraction Method: Maximum Likelihood.		

Table 1 shows that the eigenvalues in the communalities are higher in the extraction than in the initial [5]; this implies that the seven factors are valid and will be able to determine the effect of E-Systems on decision-making among the universities.

Table 2. Total Variance Explained

Factor	Initial Eigenvalues			Extraction Sums Of Squared Loadings		
	Total	% of Variance	Cumulative %	Total	% of Variance	Cumulative %
1	2.975	42.501	42.501	2.321	33.155	33.155
2	.870	12.428	54.929			
3	.821	11.725	66.654			
4	.727	10.385	77.039			
5	.613	8.761	85.799			

6	.554	7.908	93.708			
7	.440	6.292	100.000			

Extraction Method: Maximum Likelihood.

From table 2, under maximum likelihood, the total variance of the seven factors can be explained by one factor [6]. The results indicated that one factor explained 42.501% of the variability.

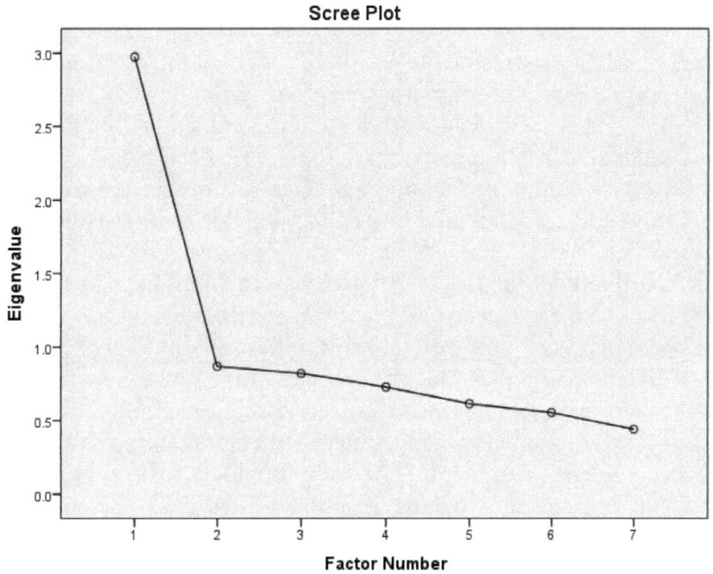

Figure 1. Scree Plot

Figure 1 shows that only one factor has an eigenvalue higher than 1, implying that one factor can account for the majority of variability in the seven factors.

Table 3.Goodness-Of-Fit Test

CHI-SQUARE	DF	SIG.
39.560	14	.000

From table 3, it can be observed that the chi square value is 39.560 with a p-value of 0.000, which is less than the 0.05 level of confidence, indicating that there is sufficient evidence that the factors can determine the effect of E-Systems on decision-making in the universities.

Table 4.Test Statistics (Chi Square Test)

	CHI-SQUARE	DF	ASYMP. SIG.
I use an e-vocational consulting programs	106.145[a]	3	.000
I use an e-extracurricular programs	113.707[a]	3	.000
I use e-vocational activities	124.717[a]	3	.000
I use an information systems	117.643[a]	3	.000
I use a computer labs to acquire knowledge	95.109[a]	3	.000
I use knowledge platforms	105.373[a]	3	.000
I use an e-exams for our vocational tendency	116.717[a]	3	.000

0 cells (0.0%) have expected frequencies less than 5.
The minimum expected cell frequency is 77.8.

IOP Conf. Series: Journal of Physics: Conf. Series **933** (2017) 012019 doi:10.1088/1742-6596/933/1/012019

From table 4, it can be observed that all the factors have p-values equal to 0.000 (very small), which is less than the 0.05 level of significance, implying that there is significant evidence that the use of an e-vocational consulting programs, use of e-extracurricular programs, use of e-vocational activities, use of informational systems, and use of e-exams for vocational tendency significantly contribute to decision-making in the universities.

4. DISCUSSION

From the results, it was observed that e-vocational consultancy was significant. The e-vocational consultancy provides significant assistance to local educators in the development and improvement of vocational education programs in different areas, such as agriculture, home economics, office trades, and technical subjects. This was achieved through conducting seminars and advising school administrators on vocational education program issues [7] , [8] . Vocational consultants work with private individuals, lawyers, courts, and public agencies to provide information about job potential, income potential, employment potential, and loss of income due to physical, emotional, educational, or employment limitations [9][10] .

The results indicated that e-extracurricular activities are significant to university decision-makers [11]. This is because most of the extracurricular activities help the decision makers to interact with the people who are in the university and get relevant information. This means that e-extracurricular activities attribute to decision-making in the universities. Employers and colleges like to see extracurricular activities on your resume, including leadership experience, athletics, foreign languages, writing/technical skills, and volunteer work, among others. Taking a leadership position while in school is a big plus on a student's resume. It shows a willingness to step up to the plate and portrays to employers that you can manage and handle responsibilities [12]. Becoming involved in campus multicultural activities can help you increase your awareness and understanding of the world's cultures and different ethnic and racial backgrounds. Such activities can include concerts, festivals, lectures, and other events that encourage multicultural awareness on campus [13]. Participating in such activities can broaden your cultural consciousness, as well as help you develop or better understand your own cultural identity.

The results found that a good knowledge platform enhances easy communication, storage of information, and analysis of easy decision making. The knowledge platforms that were found to be efficient were online platforms, as they were able to incorporate all these elements at the same time. The results show that the knowledge platform helps the decision makers to produce better results. The platform is reliable to many people because it gives all the information that is required by various participants in the effort of achieving effectiveness [14]. However, it is important to note that the results do not indicate the efficiency of the platform because there are other factors that affect performance, such as environment, knowledge management framework, and factors related to platform users [15]. The researcher observed that the participants had the same academic identities because they shared the same academic platform. The results that were posted in the study suggest that academic influences might have affected the choice and use of various knowledge platforms in doing research[16]. It is worth noting that academic influences affected the participant's attitude towards different knowledge management systems, resulting in different patterns of system utilization and performance that negatively affected the organization. Many students chose an online knowledge platform because of its enhanced features without complication and to collect information for research [17]. The availability of a knowledge platform with a friendly user interface was preferred because it created a better environment for research and easy communication. The participants were given a small number of relevant cues in graphical presentations to show the relationship between the variables and thus make it easy to make forecasts. The graphical designs helped the participants to interpret the results, thus offering better judgment as well as decision-making [18].

The e-exams have also improved a great deal to be friendlier to the students, both those in youth age brackets and those in working fields. This improvement has allowed people to grow career wise, hence making it a very good platform for decision-making [19].

5. CONCLUSION

From the research study, we can conclude that the e-system is an instrumental tool to help students in their decision-making. Most of the information lies within the e-system in all the sectors examined. In universities, decision makers can interact with the students through the E-Systems and arrive at decisions that are good for educational growth as well as growth within the job markets. The achievable results indicated that there are diverse situations due to different environments. The study further evaluated the aim of the study and discussed the literature behind the knowledge platform. The results also indicate that knowledge platforms encourage and generate improved performance and decision-making in universities. Therefore, there is need for well-established knowledge platforms to acquire the best information for decision-making.

Through E-Systems, the decision makers can make decisions on the use of e-exams as well as the knowledge platforms. The need to expand the e-system in vocational learning was brought out.

5.1 Limitations and implication

The study was limited to universities, which are just a small section of the economy in terms of producing entrepreneurial skills. This therefore hinders the use of this research finding toward other sectors of the economy. The study is also limited to students in the universities and colleges; therefore, this research work must be extended to other learning institutions.

5.2 Recommendations

The study recommends that the e-system be implemented to all areas of decision-making other that the universities. This also recommends that this research work can be important if it can evaluate the integration of E-Systems in decision making in various sectors, such as: sports, e-learning, communications, policy making, and business through e-business. As well the study recommends to study the impact of e-system of vocational guidance on future functional students' attitudes.

5.3 Future research

The study recommends to conduct further positivist research in the impact of e-system of vocational guidance on future functional students' attitudes. As well the study recommends to conduct research in the e-system in areas of security, as information tends to move faster.

6. References

[1] N. Entwistle and N. Entwistle, "Changing conceptions of learning and teaching," ENTWISTLE, N.(ed.), 1990.
[2] J. R. Fraenkel and N. E. Wallen, How to design and evaluate research in education: McGraw-Hill Higher Education, 2003.
[3] D. Gleeson, Training and its Alternatives: Open University Press Milton Keynes, 1990.
[4] H. Colley, D. James, K. Diment, and M. Tedder, "Learning as becoming in vocational education and training: class, gender and the role of vocational habitus," Journal of vocational education and training, vol. 55, pp. 471-498, 2003.
[5] C. L. Coryn and K. A. Hobson, "Using nonequivalent dependent variables to reduce internal validity threats in quasi-experiments: Rationale, history, and examples from practice," New Directions for Evaluation, vol. 2011, pp. 31-39, 2011.
[6] A. G. Bluman, Elementary statistics: A step by step approach, A brief version: New York: McGraw Hill, 2008.
[7] P. Osano and P. B. Corcoran, "Young people, education, and sustainable development: Exploring principles, perspectives, and praxis," ed: Wageningen Academic Publishers, 2009.
[8] C. Cheong, "From group-based learning to cooperative learning: A metacognitive approach to project-based group supervision," Informing Science: The International Journal of an Emerging Transdiscipline, vol. 13, pp. 73-86, 2010.
[9] C. Dahlman and D. Z. Zeng, "ENHANCING THROUGH LIFELONG LEARNING," 2007.

[10] M. M. Robles, "Executive perceptions of the top 10 soft skills needed in today's workplace," Business Communication Quarterly, vol. 75, pp. 453-465, 2012.

[11] C. J. Kirchhoff, M. Carmen Lemos, and S. Dessai, "Actionable knowledge for environmental decision making: broadening the usability of climate science," Annual Review of Environment and Resources, vol. 38, pp. 393-414, 2013.

[12] T. Wilson, "Youth unemployment: review of training for young people with low qualifications," 2013.

[13] A. L. Brown, "The advancement of learning," Educational researcher, vol. 23, pp. 4-12, 1994.

[14] D. R. Anderson, D. J. Sweeney, T. A. Williams, J. D. Camm, and J. J. Cochran, An introduction to management science: quantitative approaches to decision making: Cengage learning, 2015.

[15] M. D. Gall, W. R. Borg, and J. P. Gall, Educational research: An introduction: Longman Publishing, 1996.

[16] M.-P. Jime´ nez-Aleixandre, "Knowledge producers or knowledge consumers? Argumentation and decision making about environmental management," International Journal of Science Education, vol. 24, pp. 1171-1190, 2002.

[17] M. N. Riaz and M. T. Khalili, "Transformational, transactional leadership and rational decision making in services providing organizations: Moderating role of knowledge management processes," Pakistan journal of commerce and social sciences, vol. 8, pp. 355-364, 2014.

[18] J. F. Courtney, "Decision making and knowledge management in inquiring organizations: toward a new decision-making paradigm for DSS," Decision Support Systems, vol. 31, pp. 17-38, 2001.

[19] J. Eison, "Using active learning instructional strategies to create excitement and enhance learning," Jurnal Pendidikantentang Strategi Pembelajaran Aktif (Active Learning) Books, vol. 2, pp. 1-10, 2010.

An Attribute Based Access Control Framework for Healthcare System

Majid Afshar[1], Saeed Samet[2], Ting Hu[3]

[1]Memorial University of Newfoundland, Department of Computer Science, St. John's, NL, Canada.
[2]University of Windsor, School of Computer Science, Windsor, ON, Canada (Adjunct Professor at Memorial University of Newfoundland, Faculty of Medicine, St. John's, NL, Canada)[3] Memorial University of Newfoundland, Department of Computer Science, St. John's, NL, Canada.

Corresponding author's email: mman23@mun.ca

Abstract. Nowadays, access control is an indispensable part of the Personal Health Record and supplies for its confidentiality by enforcing policies and rules to ensure that only authorized users gain access to requested resources in the system. In other words, the access control means protecting patient privacy in healthcare systems. Attribute-Based Access Control (ABAC) is a new access control model that can be used instead of other traditional types of access control such as Discretionary Access Control, Mandatory Access Control, and Role-Based Access Control. During last five years ABAC has shown some applications in both recent academic fields and industry purposes. ABAC by using user's attributes and resources, makes a decision according to an access request. In this paper, we propose an ABAC framework for healthcare system. We use the engine of ABAC for rendering and enforcing healthcare policies. Moreover, we handle emergency situations in this framework.

1. Introduction

The principal concern of Personal Health Records (PHR) is confidentiality [1], especially when patients use commercial healthcare distributed based systems, such as Microsoft HealthVault [2] or WebMD, to maintain their health records. Access Control Mechanism (ACM) is useful for providing access policy for different types of users. However, traditional access control mechanisms, such as Role-Based Access Control (RBAC) and Identity Based Access Control (IBAC), have several restrictions in distributed based systems [3], [4]. In particular, the data owner is unable to control the information when they upload their data to the server.

Access control means the process of defining and managing which users are permitted to perform what operations on which objects in a system. Availability of resources is one of the aspects of access control that conflicts with restricting access to resources that is another perspective of access control [5]. Access control creators need to know how the organization functions work and the access requirements of the system users. In case of the high-level user restrictions, the result has at least two kinds of disadvantages. First, if the user with a valid request cannot access to seeking information, then she tries to find other ways to accessing the required recourses. Second, and more important in healthcare systems is the lack of exception case handling.

Several of the access control mechanisms have been proposed in healthcare information systems [6], [7], [8], [9] without addressing exception or emergency handling. However, some of those have had to pass exceptions to the typical access control mechanisms to be able to satisfy the needs of their users in emergency cases. It is evident that handling of exception cases is not easy due to its effect on control over information flow in a healthcare system. The purpose of this study is to apply the concept of ABAC instead of existing current access control mechanisms, and also to address the problem of emergency cases in healthcare applications.
The rest of this paper is organized as follows. Section 2 includes two subsections on background and related work. In section 3, we show an ABAC framework for the healthcare system. Section 4 describes how to manage emergency situations. In section 5, contains conclusions and future work.

2. Background and Related Works

2.1. Background
ABAC describes an access control model where access rights are admitted to users by the use of policies [4]. Attributes are properties that describe specific features of the subject, object, environment conditions, and requested actions that are predefined and preassigned by an owner or administrator or authority. Attributes are formed of an arbitrary section that indicates the class of information supplied by the attribute, a name, and value.

A subject is an actual entity such as a person, process, or device that by sending requests makes changes to the system's status. It could be a user, requestor, or mechanism acting on behalf of a user or requestor. In this work, we assume that subject and user have the same meaning. An object is a passive system-related entity that contains or receives information such as files, records in a table, databases, applications, networks, and domains. For the goal of this work, we assume that object and resource are synonymous.

By processing of a subject's request upon an object, the system runs a particular function, which is called an operation. Operations could be one or more of the following functions: read, write, edit, delete, author, copy, execute, and modify. The policy is the description of rules or connections that determine the set of permissible operations a subject should run upon an object in authorized environment conditions. The policies could contain the combination of the following kinds of attributes [4], [10]:

User (subject) attributes, resource (object) Attributes, environmental attributes (conditions) and administrativeattributes.

*2.2. RelatedWork*The Role-Based Access Control (RBAC) model was formally presented by David F.Ferraiolo and Richard Kuhn [11]. By using RBAC, organizations can assign users to defined role(s). Then users can obtain all the permissions of the roles of which they are the member. Roles can be generated for the various tasks in an organization and based on ability, and eligibility, users are assigned to the roles. Users are able to select other roles based on the system policy management. Roles can be given new permissions from resources, and permissions can be revoked from roles as required.

The above description has turned RBAC as a primary model of option for access control applications. However, RBAC suffers from some major problems, especially in healthcare system. First and the major weakness of RBAC is role explosion [12]. When an organization that using RBAC faces a rising number of various real-world roles it needs an increasing quantity of RBAC roles to define permissions well. In this case, handling all those roles becomes a complicated circumstance[13]. Next, RBAC gives an almost same level of access to users who have the same roles. Although RBAC is scalable and adaptable and has been implemented a lot in different applications [14], [15], [16], it is not proper for the healthcare systems [17].

3. Access Control Architecture

As shown in Fig 1, this example contains the following primary functional nodes: The Policy Enforcement Point (PEP), the Policy Decision Point (PDP), the Policy Administration Point (PAP) and the Policy Information Point (PIP). Those parts also called ABAC engine. Once aforementioned segments are in a schema together, they have to perform in compatibility to present proper access control decisions.

As part of authorization process, PDP is the heart of the system and responsible for assessment on policies and attributes that come from PAP and PIP, respectively, to return an access control decision. In other words, it makes the access decisions by assessing the appropriate policies and attributes. The PDP should return the decision procedures based on the ACM's computational languages like XACML [18]. Of course, one of the principal duties of the PDP is to reduce the risk of collision between system polices and corresponding Meta-Polices (MP).

Then, the output of PDP must be enforced by the next component. This function refers to the PEP. In response to a user's request, the PEP performs the appropriate access decisions that is prepared by the PDP. In this system, user could be a doctor, researcher or nurse. The result or output of PEP could either allow or deny user access to the requested protected patients' records.

Packages of PDP and PEP work as authorization services and they need some information regarding user and patient's attributes as part of the input to properly perform their duties. PIP addresses this issue, and it works as the retrieval source of attributes as well as healthcare environmental data, and other data needed for policy assessment. For storing and retrieving polices in an appropriate repository, we need another function, called PAP. PAP should also be able for creating, editing, running and debugging policies and MPs by a user interface. To explain how this proposed framework can handle a user request and protect patient's privacy, we use a simple scenario as follows.

First of all, we should have the primary policy for the whole system. Consider in a hospital, doctors could have access to all patient's individual records, and nurses just have access to identity patients' information, and researchers can see statistical information only if the number of patients is greater than 50. Besides, each set of policy must have a supplementary segment that specifies requester, type of resource and action. In this case, data requester must be a doctor, nurse or researcher. Resource demonstrates clinical health data like patient's individual records that contain high-level sensitive data, and the action is the set of proper functions that data requester is authorized and permitted to do on the resource, such as read, write and execute. If all conditions mentioned in supplementary segment of the policy are satisfied by a data requester, like a doctor, then the policy set might be considered as an applicable policy.

Consider the following table that shows our example and some additional conditions in supplementary segment. Each employee in hospital has a role attribute. If the value of this role equals to doctor, nurse or researcher, then they might be able to see some results from sensitive health information.

Assume a doctor as a data requester needs to read a particular patient's record and this record has been registered two years ago (Request 1). Based on the supplementary segment in Table 1 that request satisfies all the conditions. Thus, the policy can consider as an applicable policy. On the other hand, patients can totally determine who can access to their health data and for what part of data by adding some policies. For instance, a patient just allows her doctor to have access only to her last year treatments. Also, the patient is not willing to share his health data with any researcher.

Based on the aforementioned patient policy (conditions), the Request 1 should be rejected due to the value of attributes of data requester, that cannot satisfy the patient's policy. Of course, the system must have a strategy in especial occasions to avoid conflicts. As an example, if the system receives two decisions that made by PDP at the same time on a same resource, then the system should call a particular algorithm to handle the conflict.

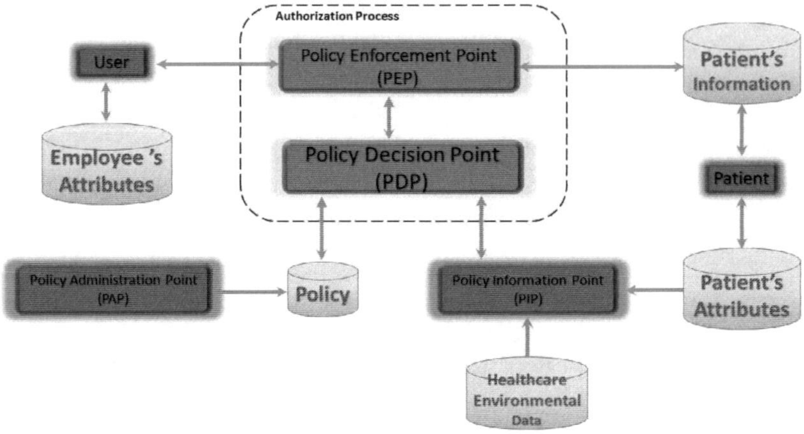

Figure 1. Using ABAC in healthcare system.

Table 1. The Main Policy Set For Hospital

Data Requester	Type of Resource	Action
Doctor	Patient's history records	Read, Write
Nurse	Identity information of patients	Read
Researcher	Statistical Information	Execute if Patient count> 50
Supplementary Segment		
Doctor	Patient's history records (only last 5 years)	Read, Write
Nurse	Identity information of patients	Read
Researcher	Statistical Information (all years expect two last years)	Execute if Patient count> 50

4. Manage Emergency Cases

Handling emergency situations in a healthcare system is so complicated. Consider a particular patient comes with or without an ambulance to the hospital, and there is no time for patient registering process that is essential for access control. We propose a mechanism to manage that situation called emergency access. In this case, the healthcare system provides some access to the patient's data. This grant is not only directly related to a job position or special role, but is also intended as a special permission associated directly with the particular data requester to be able to satisfy all the following conditions with her attributes.

- Data requester must be a doctor.
- The doctor must work in an emergency section.
- The doctor must be at work at the time of the request.

The reasoning for this policy is that only a limited number of data requesters might be able to assign this permission, and they must be decided on their attributes not just based on the job position. Besides, there is a possibility of tracing the usage of emergency access by an administrator. Normally only a few users at emergency sections might be able to gain this permission. Although emergency access is only supposed to be applied in emergency cases, to prevent any abuse of this grant, we appended some additional conditions after assigning the permission. The conditions may modify lightly from one system to another, but the most common ones can define as follows:

- If the patient sets some policies after a while, then emergency access is not valid any more.
- By using the emergency access just permit user to have access only to a subset of the available patient information, not the complete patient record. To gain access to the other parts of the record, the user has to pass some other system policies.

- Information obtained using emergency access will only remain accessible for the data requester for a short period of time that can be defined based on average time needed.
- Regaining access to the patient's information after a timeout is needed to satisfy all the conditions again.

5. Conclusion

In this paper, we have illustrated the application of ABAC as a framework in the healthcare systems. Besides, we have discussed on how this framework can manage and enforce policies to grant or deny data requesters based on their attributes. ABAC supports accommodating the unexpected users. By using this feature and applying some conditions as an emergency access policy, we have demonstrated how an emergency situation could be handled with patient's privacy protection.
Although ABAC has several advantages in the healthcare systems, it is still largely in its infancy, and the list of future work and open problems are extensive. For instance, we could add some other features to address other real scenarios. Implementing the new framework with new features and check the corresponding performance are among other future directions of this work.

6. References

[1] Señor, Inmaculada Carrión, José Luis Fernández-Alemán, and Ambrosio Toval. "Are personal health records safe? A review of free web-accessible personal health record privacy policies." Journal of Medical Internet Research14.4 (2012).
[2] Sunyaev, Ali, Dmitry Chornyi, Christian Mauro, and Helmut Krcmar. "Evaluation framework for personal health records: Microsoft HealthVault vs. Google Health." In System Sciences (HICSS), 2010 43rd Hawaii International Conference on, pp. 1-10. IEEE, 2010.
[3] Kalam, Anas Abou El, R. El Baida, Philippe Balbiani, Salem Benferhat, Frédéric Cuppens, Yves Deswarte, Alexandre Miege, Claire Saurel, and Gilles Trouessin. "Organization based access control." In Policies for Distributed Systems and Networks, 2003. Proceedings. POLICY 2003. IEEE 4th International Workshop on, pp. 120-131. IEEE, 2003.
[4] Hu, Vincent C., David Ferraiolo, Rick Kuhn, Arthur R. Friedman, Alan J. Lang, Margaret M. Cogdell, Adam Schnitzer, Kenneth Sandlin, Robert Miller, and Karen Scarfone. "Guide to attribute based access control (ABAC) definition and considerations (draft)." NIST special publication 800, no. 162 (2013).
[5] Fisler, Kathi, Shriram Krishnamurthi, Leo A. Meyerovich, and Michael Carl Tschantz. "Verification and change-impact analysis of access-control policies." In Software Engineering, 2005. ICSE 2005. Proceedings. 27th International Conference on, pp. 196-205. IEEE, 2005.
[6] Blobel, Bernd. "Trustworthiness in distributed electronic healthcare records-basis for shared care." In Computer Security Applications Conference, 2001. ACSAC 2001. Proceedings 17th Annual, pp. 433-441. IEEE, 2001.
[7] Blobel, Bernd. "Authorisation and access control for electronic health record systems." International journal of medical informatics 73, no. 3 (2004): 251-257.
[8] Hu, Junzhe, and Alfred C. Weaver. "A dynamic, context-aware security infrastructure for distributed healthcare applications." In Proceedings of the first workshop on pervasive privacy security, privacy, and trust, pp. 1-8. 2004.
[9] Schwartmann, Dirk. "An attributable role-based access control for healthcare." In International Conference on Computational Science, pp. 1148-1155. Springer, Berlin, Heidelberg, 2004.
[10] Servos, Daniel, and Sylvia L. Osborn. "Current Research and Open Problems in Attribute-Based Access Control." ACM Computing Surveys (CSUR) 49, no. 4 (2017): 65.
[11] David, Ferraiolo, and Kuhn Richard. "Role-based access controls." In Proceedings of 15th NIST-NCSC National Computer Security Conference, vol. 563. Baltimore, Maryland: NIST-NCSC, 1992.
[12] Elliott, Aaron, and Scott Knight. "Role Explosion: Acknowledging the Problem." In Software

Engineering Research and Practice, pp. 349-355. 2010.
[13] Sandhu, Ravi, Venkata Bhamidipati, and Qamar Munawer. "The ARBAC97 model for role-based administration of roles." ACM Transactions on Information and System Security (TISSEC) 2, no. 1 (1999): 105-135.
[14] Tang, Bo, Qi Li, and Ravi Sandhu. "A multi-tenant RBAC model for collaborative cloud services." In Privacy, Security and Trust (PST), 2013 Eleventh Annual International Conference on, pp. 229-238. IEEE, 2013.
[15] Li, Wei, Haishan Wan, Xunyi Ren, and Sheng Li. "A refined RBAC model for cloud computing." In Computer and Information Science (ICIS), 2012 IEEE/ACIS 11th International Conference on, pp. 43-48. IEEE, 2012.
[16] Takabi, Hassan, James BD Joshi, and Gail-Joon Ahn. "Security and privacy challenges in cloud computing environments." IEEE Security & Privacy 8, no. 6 (2010): 24-31.
[17] Mukherjee, Subhojeet, Indrakshi Ray, Indrajit Ray, Hossein Shirazi, Toan Ong, and Michael G. Kahn. "Attribute Based Access Control for Healthcare Resources." In Proceedings of the 2nd ACM Workshop on Attribute-Based Access Control, pp. 29-40. ACM, 2017.
[18] Rissanen, Erik. "Extensible access control markup language (xacml) version 3.0." (2013).

Acknowledgments

This work has been partially supported by the Natural Sciences and Engineering Research Council of Canada (NSERC) and the Research & Development Corporation of Newfoundland and Labrador (RDC).